Complexity and Emergence in Market Ecosystems

New Economic Windows

Debora Sarno · Francesco Polese ·
Stephen Louis Vargo
Editors

Complexity and Emergence in Market Ecosystems

Springer

Editors
Debora Sarno
Department of Management
and Quantitative Studies/DISAQ
Parthenope University of Naples
Naples, Italy

Francesco Polese
Department of Political Sciences
and Communication/DISPC
University of Salerno
Salerno, Italy

Stephen Louis Vargo
Department of Marketing and Supply Chain
Management
University of Oklahoma
Norman, OK, USA

ISSN 2039-411X ISSN 2039-4128 (electronic)
New Economic Windows
ISBN 978-3-031-81944-5 ISBN 978-3-031-81942-1 (eBook)
https://doi.org/10.1007/978-3-031-81942-1

This Springer imprint is published by the registered company Springer Nature Switzerland AG
The registered company address is: Gewerbestrasse 11, 6330 Cham, Switzerland

Foreword

This book is timely because a phase change is at hand. A phase change is a big, noticeable change. In the natural world, people are familiar with phase changes, such as water turning to steam, or water turning to ice; this type of phase change provides a routine service like a morning espresso or gelato for dessert. However, the diverse phase changes coming to market ecosystems that are described in the pages ahead are likely to impact service in far more profound ways.

The amazing complexity of our human-made world is revealed in each chapter of this engaging scholarly book. Each chapter ahead provides a unique perspective, along with an insight, a warning, a prediction, a hope, or a dream of what might be in store for humanity as new technological capabilities create pressures that challenge institutions to keep up with amplified good, bad, and accidental actors at multiple scales. Because of centuries of advances in the physical sciences, businesses have learned to control a wide range of phase changes in nature to serve a purpose. As the market power of some businesses grows (as measured by market capitalizations in the trillions of dollars), governments are scrambling and struggling to co-create institutions with all stakeholders to shape the phase changes on the horizon.

This book builds on the shoulders of giants. Systems thinkers like von Bertalanffy, Beer, Simon, Golinelli, and many others are referenced in the chapters ahead. Indeed, readers of this book are most likely to be systems thinkers themselves—whether in academics, industry, or government. And yet, it is not easy to communicate about systems across departmental disciplinary, industry sector, and cultural boundaries. Will advancing technologies and institutional needs lead to a phase change in our abilities to communicate with each other? Authors throughout the centuries have struggled to make what is clear to them, explicit and clear in the minds of their readers. Will we remain storytellers gathered around the campfire, faces illuminated by flickering flames or LED screens while blindly scrolling, or shape a phase change in which self-control and technology are simultaneously harnessed to amplify capabilities, scale up benefits, and mitigate harms?

This book will challenge all readers to rethink their world from a Service-Dominant (S-D) logic perspective. Most readers would agree that historically the transition from child to adult, from illiterate to literate, from rural to urban, from in-person to online are profound in terms of rethinking what activities we do and how we do them on a day-to-day basis. Equally profound, but less appreciated today, is the mind shift that comes from adopting an S-D logic perspective. Why do I make this observation after reading a draft of this book? Not simply because the chapters make many references to S-D logic but because of something more fundamental. Seeing the world in terms of multi-scale responsible actors, such as people, businesses, and nations, that exist to interact and change, by giving service, getting service, and striving to grow to become better future versions of themselves in the give-get-grow of service, is in fact fundamentally the story of civilization and progress in a nutshell.

Will anything remain the same after the phase change ahead? Yes, and it is service interaction and change all the way down and up. In nature, the evolving ecology of species, before humans existed, can be viewed as the story of more and more complex win-win interaction and change. I know many will say this is not true, instead it is dog eat dog and survival of the most adaptable to change, but I challenge them to a debate on this point, after they read this book. The long arc of history is surprising, and readers of this book, will surely get a glimpse of the importance of S-D logic in shaping the views of the chapter authors and myself as well. While value co-creation and value co-destruction can both occur, one is more sustainable than the other, and systems can "learn" amazing things, given enough time.

The chapters of this book are a series of steppingstones into the future. Let me explain what I mean with a story. Earlier this year, my oldest son and I were hiking a somewhat challenging trail at the back entrance of Henry Coe State Park in California where we live and enjoy exploring nature together these days. The trail was in a long valley, high hills on both sides, and a rapidly flowing stream crisscrossed our trail repeatedly, with only a few stones to balance on as we crossed each time. At first, it was easy to slowly step across stone by stone because many people had hiked in that far, and each hiker had added a stone building an easy walkway ahead. However, as we hiked further and further into the wilderness valley, there were fewer steppingstones at each crossing. Instead of easily walking across, we had to take some risk and run across hopping from stone to stone hoping we did not get a drenching in the rapidly flowing creek. Finally, we had gone so far, we had to search for big stones to throw in the water ahead and make our own steppingstone pathway.

The authors of each of the chapters of this book, in a way, are like my son and I working to build safe steppingstones to get to the other side. However, in the case of the chapter authors, the other side is the other side of the coming phase change. Some chapters feel well-traveled to us systems thinkers. Some chapters we hurry across, hoping not to get drenched. Other chapters demand we co-create with the author and look for steppingstones of our own to throw into the rapidly flow streams of ideas.

This book is timely because a phase change is at hand. Chapter 1 introduces phase transitions in service ecosystems. Chapter 2 uses a systems perspective to reconsider market shaping by actors; responsible actors learning to invest wisely in shaping markets in a world we would want. Chapter 3 runs ahead even faster, exploring innovation, resilience, and sustainability. Chapter 4 enters the realm of complexification. Chapter 5 gets us to focus on cities. Chapter 6 entices us to think about open platforms. Chapter 7 provokes us to take a hard look at higher education—ouch, but so important. Chapter 8 could only be written by true visionaries who can see into the future realms. Chapter 9 challenges us to step through the looking glass and go into one possible, shocking future. Chapter 10 reels us back into what is truly important about traditions even as we go digital. Chapter 11 explores the evolution of emotional intelligence in the digital age. As a species, with capabilities amplified by our clever tricks and technologies, can we shape this next phase change with enough self-control of heart, mind, and will in order to truly move beyond stories around the campfire?

As the sun sets on another day, I see in this book 11 steppingstones leading us into the future. Each steppingstone is surrounded by a rapidly flowing stream of world events and technological advances. However, what my eyes see most of all is the glistening flow, catching a few sparkles of sunlight—the give, get, grow of service—a reassuring constant in an ever-changing world. I hope you enjoy reading this book as much as I did.

Morgan Hill, CA, USA
October 2024

Jim Spohrer
Retired Industry Executive (IBM and Apple), Board Member (International Society of Service Innovation Professionals and ServCollab), and Senior Fellow (University-Industry Development Program)

Preface

“No man is an island,” “no actor is an island”... have been recurrent themes in the literatures of diverse disciplines, from religion to marketing. Similarly, “nothing happens in isolation” has been mentioned very frequently in physics, as well as in history. The central thesis of these views is that every actor’s interaction, every exchange, every learning experience, in other words every action, happens in context of dynamic systems, made up of diverse, interrelated actors, resources, relationships, and structures, that are always changing and adapting. Understanding these dynamics has been manifested in what has been called a *systems turn* in the marketing and management (and other) literatures during the last century. Although the systems’ door has been opened, the universe of theoretical conceptualization and managerial (and other) implications of looking at the world through a system’s lens has still not been fully grasped. Opening the systems’ door further reveals a number of essential concepts that need to be more fully investigated and understood if this turn is to be extended.

Complexity and *emergence* are two of the most central of these concepts, ones with which we, the co-editors of this volume, have long been intrigued. *Systems thinking* and, more generally, systems theories, have been the focus of a significant segment of the Italian school of management[1]—with which Debora and Francesco have long been associated (the *viable systems approach*)—which adopts a holistic, systemic perspective on firms’ organization, relationships, and interactions. *Service dominant logic* is another initiative—co-founded by a third editor—which has leveraged systems thinking and complexity theory to develop a theoretical framework, based on a service ecosystem perspective.[2] It was originally developed to make sense

[1] It has generated the Viable Systems Approach. See Golinelli, G. M. (2010). Viable systems approach (VSA). Governing business dynamics. Cedam; Barile, S. & Polese, F. (2010). Linking the Viable System and many-to-many network approaches to Service Dominant logic and Service Science. *International Journal of Quality and Service Sciences*, 2(1), 23–44; Barile, S. (2024). *Systems Thinking and Viable Systems*. Routledge.

[2] Vargo, S. L. & Lusch, R. F. (2016). Institutions and axioms: an extension and update of Service-Dominant logic. *Journal of the Academy of marketing Science*, *44*(1), 5–23; Vargo, S. L., and R. F. Lusch. *The SAGE handbook of Service-Dominant logic*. London: Sage Publications.

of value co-creation among actors in markets, but which has later found resonance in other contexts and disciplines and has been used foundationally in *service science*.[3] Complexity is the leitmotiv of those theoretical frameworks, and emergent properties (unpredictable and irreducible to the parts generating them) are essential outcomes of actors' interaction in complex adaptive systems.

Marketing, service, and related scholars from around the world, driven both by academic concerns and the challenges of contemporary economic and social systems for managers and other concerned parties, have been meeting for the last decade or two to discuss complexity and emergence. Two venues for these discussions have been the *Naples Forum on Service (NFS)*[4] and of the *Forum on Markets and Marketing (FMM)*[5], which are held biennially, in alternate years. Multiple papers from scholars participating in these forums have contributed to building a wider interest in the topics among other scholars in diverse disciplines. Perhaps of particular note was the *Massa Lubrense Emergence Forum* of 2019, a pre-meeting of NFS involving scholars with extensive systems' reading backgrounds from diverse (sub)disciplines. During the 3-days meeting, held in a beautiful location on the sea with a view on Capri, the storyline and contribution of what became a 2023 article on emergence, published in the *Journal of the Academy of Marketing Science,* was developed.[6]

A third term, related to complexity and emergence, that has surfaced in these discussions, is *phase transitions*, large-scale, step-changes of systems.[7] In addition to their role in complex adaptive systems in general, interest in phase transition was likely also motivated by recent, real-world large-scale changes brought on by perturbations such as the COVID-19 pandemic, as well as the dramatic manifestations of climate change, leading to, if not necessitating, various social, energy[8] and other transitions. Regardless of the catalyst, the concept of transition is now a central focus of systems thinking.

Increasing recognition of all of these and other concepts has led us, joined by colleagues and other interested scholars, to explore them further. Some of the results of these discussions have already appeared on important journals; others have found their home in this book, co-created with a group of other 27 passionate academics and practitioners willing to share, and receive feedback on, the advancements of their research. However, much of the hard work of more deeply drilling down on and operationalizing those concepts remain. Although the still nascent systems' turn

---

[3] Maglio, P. P., & Spohrer, J. (2008). Fundamentals of service science. *Journal of the academy of marketing science*, *36*, 18–20.

[4] naplesforumonservice.com

[5] sdlogic.net

[6] Vargo, S. L., Peters, L., Kjellberg, H., Koskela-Huotari, K., Nenonen, S., Polese, F., Sarno, D. & Vaughan, C. (2023). Emergence in marketing: an institutional and eco-system framework. *Journal of the Academy of Marketing Science*, 51(1), 2–22.

[7] Polese, F., Payne, A., Frow, P., Sarno, D., & Nenonen, S. (2021). Emergence and phase transitions in service ecosystems. *Journal of Business Research*, 127, 25–34.

[8] Sarno, D., & Siano, P. (2022). Exploring the Adoption of Service-Dominant Logic as an Integrative Framework for Assessing Energy Transitions, *Sustainability*, vol. 14(15), 1–26, August.

has begun to manifest, a phase transition of its own still requires more diffusion, understanding, and adoption by additional researchers and practitioners. We thank both the authors and readers of this effort for their participation and we hope it will move is closer to that goal.

Naples, Italy
Salerno, Italy
Norman, USA
October 2024

Debora Sarno
Francesco Polese
Stephen Louis Vargo

About This Book

The world we live in is characterized by turbulence, unpredictability, and wicked problems. In such a context, managers, public policymakers, and others are trying to better understand how foster favorable changes. This book deals with the complexity of contexts and the change processes, particularly the ones radical and disruptive that can drive markets, social systems, organizations, cities, and other service ecosystems towards a new order and configuration. This book presents frameworks, typologies, other schemas, and insights from diverse theories (e.g., complexity theory, emergence theory, and service-dominant logic), to support academics and practitioners in analyzing, understanding, and try to envisioning transformations of their systems, including those leading to better sustainability.

The work is the result of research and discussions shared during international academic events such as the Naples Forum on Service and the Forum of Markets and Marketing over the last 5 years and is enriched by contributions from the network of the colleagues involved in those discussions. The 11 chapters are the result of studies that can be seen as ranging in diverse levels of aggregation (numbers of actors involved) and abstraction (specificity/generalizability of related concepts),[9] as shown in Fig. 1.

Chapter 1, "Exploring Radical Market Changes as Phase Transitions of Service Ecosystems: Insights from Complexity Science," by Debora Sarno, Jaqueline Pels, Francesco Polese, David Sörhammar, Bård Tronvoll, and Stephen L. Vargo, explores how complexity science can contribute to the understanding of phase transitions in service ecosystems. In particular, three diverse approaches to them are deepened and reconciled. Illustrations are related to markets in transitions due to digitalization. The chapter highlights several topics within complexity science that can be further investigated within service-dominant logic.

Chapter 2, "Using a Systems Perspective for Sharing Market Views and Changes for Market Shaping," by Debora Sarno, Francesco Polese, and Antonietta Megaro, focuses on systems theories and the concept of change to contribute to market shaping

[9] Vargo, S. L., & Lusch, R. F. (2017). Service-dominant logic 2025. *International journal of research in marketing, 34*(1), 46–67.

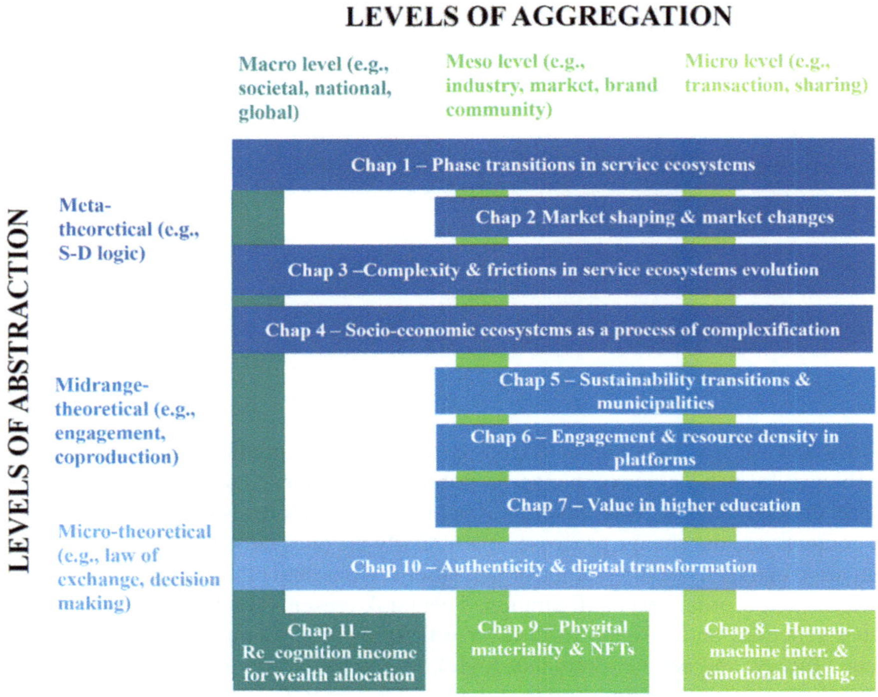

Fig. 1 Main focus of the 11 chapters of this book based on levels of abstraction and aggregation. *Source* framework adapted from Vargo and Lusch, 2017

as a collective action. The chapter reconciles the diverse viewpoints on actors by delineating what should be shared among shapers. Then, the types of market changes are differentiated to enable shapers to recognize the extent of change (or change type) they may instill in the market and the possible outcomes in terms of the degree of predictability and manageability of the systemic outcomes of shaping.

Chapter 3, "Complexity, Frictions, and Evolution in Service Ecosystems," by Alessia Anzivino, Debora Sarno, and Roberta Sebastiani, explores the complexity of a service ecosystem as perceived by actors, distinguishing among diverse complexity states. Then, it relates such states to the main frictions experienced by actors highlighting their role in the evolution of service ecosystem. Illustrations on healthcare systems pervaded by technology are presented.

Chapter 4, "Socio-economic Ecosystems Seen as Process of Complexification," by Petter Braathen, focuses on the multi-level and hierarchical complexity of systems, adopting an evolutionary-systemic and interdisciplinary perspective. Adaptation and co-evolution of socio-economic ecosystems is related to a process of complexification and emergence of collective dynamics.

Chapter 5 "New Approaches to Steer Sustainability Transitions Within Service Ecosystems in the Context of Municipalities," by Anna Viljakainen, Kirsi Maria

Hyytinen, and Katri Valkokari, transitions are investigated with a specific focus on the possible sustainability impact within municipalities. A systemic framework co-created with a municipality is shown to identify strategies for reorganization, highlighting the impact on social, technical, and economic spheres.

Chapter 6, "The Innovation Effects of Engaged Actors in Open Platform Ecosystems," by María J. Quero and Marco Luzi, deals with emergence of platforms—especially rewards-based crowdfunding open platforms—considering them as service ecosystems. Thus, it deals with the decisions related to resource density and potential for innovation, mediated by actors' engagement. A strategy matrix is proposed to offer actors actionable innovation strategies for diverse levels of actors' engagement and resource density.

Chapter 7, "The Value Depletion Circle in Higher Education and Its Evolution to Nowhere: SDL Rescue," by Montserrat Díaz-Méndez, Michael Saren, Moisés Zamora-Ramos, and Antonio Chamorro Mera, discusses value co-creation and viability of societies when privatization practices worth for business are imported in higher education service ecosystems.

Chapter 8, "The Complexity of Interactions Between Automation and Human in Service Provisioning. The Role of Emotional Intelligence," by Marisa Faggini, Marzia Del Prete, Bruna Bruno, and Anna Parziale, deals with the complexity of human–machine interactions, considering the emotional intelligence of AI as a tool to improve customer engagement. The chapter supports the logic of human–machine complement rather than substitution of human operators by AI.

Chapter 9, "Phygital Materiality in Service Ecosystems: The Transformative Role of NFTs," by Tiziana Russo-Spena, Cristina Mele, and Stefano Paolo Russo, explores the complex scenarios for service ecosystems—and their resources, actors, and value co-creation processes—that can be transformed due to the introduction of NFTs.

Chapter 10, "Navigating Digital Transformation in Cultural Organizations: Balancing Authenticity, Tradition, and Innovation," by Saba Samadilashkariania and Jon Engströmb, copes with the perception of authenticity as a property emerging out of actors interactions and ne gotiations. The topic is particularly relevant when dealing with the introduction of digitalization in traditional contexts, such as the one analyzed in the chapter (the Royal Swedish Opera).

Chapter 11, "Towards a 'Re_cognition Income': Addressing Wealth Reallocation in the Digital Age," by Sergio Barile and Francesca Iandolo, introduces the emergence of participatory capitalism driven by the recent and rapid technological advancements and discusses its impact on wealth distribution. It proposes the concept of 'Recognition Income' as a solution for equitable wealth reallocation, sharing the benefits of technological advancements, social justice, and sustainability.

Contents

1 **Exploring Radical Market Changes as Phase Transitions of Service Ecosystems: Insights from Complexity Science** 1
Debora Sarno, Jaqueline Pels, Francesco Polese, David Sörhammar, Bård Tronvoll, and Stephen Louis Vargo

2 **Using a Systems Perspective for Sharing Market Views and Changes for Market Shaping** 29
Debora Sarno, Francesco Polese, and Antonietta Megaro

3 **Complexity, Frictions, and Evolution in Service Ecosystems** 55
Alessia Anzivino, Debora Sarno, and Roberta Sebastiani

4 **Socio-Economic Ecosystems Seen as Process of Complexification** 75
Petter Braathen

5 **New Approaches to Steer Sustainability Transitions Within Service Ecosystems in the Context of Municipalities** 95
Anna Viljakainen, Kirsi Maria Hyytinen, and Katri Valkokari

6 **The Innovation Effects of Engaged Actors in Open Platform Ecosystems** 113
María J. Quero and Marco Luzi

7 **The Value Depletion Circle in Higher Education and Its Evolution to Nowhere: SDL Rescue** 129
Montserrat Díaz-Méndez, Michael Saren, Moisés Rubén Zamora-Ramos, and Antonio Chamorro-Mera

8 **The Complexity of Interactions Between Automation and Human in Service Provisioning. The Role of Emotional Intelligence** 149
Marisa Faggini, Marzia Del Prete, Bruna Bruno, and Anna Parziale

9 **Phygital Materiality in Service Ecosystems: The Transformative Role of NFTs** 167
Cristina Mele, Tiziana Russo Spena, and Stefano Paolo Russo

10 **Navigating Digital Transformation in Cultural Organizations: Balancing Authenticity, Tradition, and Innovation** 187
Saba Samadilashkariani and Jon Engström

11 **Towards a 'Re_cognition Income': Addressing Wealth Reallocation in the Digital Age** 211
Sergio Barile and Francesca Iandolo

Contributors

Alessia Anzivino Università Cattolica del Sacro Cuore, Milan, Italy

Sergio Barile Department of Management, Sapienza University of Rome, Rome, Italy

Petter Braathen Inland Norway University, Kongsvinger, Norway

Bruna Bruno Associate Professor in Macroeconomics, Department of Economics and Statistics, University of Salerno, Salerno, Italy

Antonio Chamorro-Mera University of Extremadura, Badajoz, Spain

Marzia Del Prete AI & Data, Deloitte (Italy), Milano, Italy

Montserrat Díaz-Méndez University of Extremadura, Badajoz, Spain

Jon Engström Stockholm Business School, Stockholm University, Stockholm, Sweden

Marisa Faggini Associate Professor in Macroeconomics, Department of Economics and Statistics, University of Salerno, Salerno, Italy

Kirsi Maria Hyytinen VTT Technical Research Centre of Finland, Espoo, Finland

Francesca Iandolo Department of Management, Sapienza University of Rome, Rome, Italy

Marco Luzi University of Málaga, Málaga, Spain;
University of Agder, Agder, Norway

Antonietta Megaro University of Salerno, Fisciano, Italy

Cristina Mele University of Naples Federico II, Naples, Italy

Anna Parziale Fellow in Macroeconomics, Department of Economics and Statistics, University of Salerno, Salerno, Italy

Jaqueline Pels Universidad Torcuato Di Tella, Buenos Aires, Argentina

Francesco Polese University of Salerno, Fisciano, Italy

María J. Quero University of Málaga, Málaga, Spain

Stefano Paolo Russo University of Naples Federico II, Naples, Italy

Saba Samadilashkariani Stockholm Business School, Stockholm University, Stockholm, Sweden

Michael Saren University of Birmingham, University of Leicester (UK), Leicester, UK

Debora Sarno University of Naples Parthenope, Naples, Italy

Roberta Sebastiani Università Cattolica del Sacro Cuore, Milan, Italy

David Sörhammar Inland Norway University of Applied Sciences, Rena, Norway

Tiziana Russo Spena University of Naples Federico II, Naples, Italy

Bård Tronvoll Inland Norway University of Applied Sciences, Rena, Norway

Katri Valkokari VTT Technical Research Centre of Finland, Espoo, Finland

Stephen Louis Vargo University of Oklahoma, Norman, US

Anna Viljakainen Aalto University School of Arts, Design and Architecture, Helsinki, Finland

Moisés Rubén Zamora-Ramos Autonomous University of the State of Hidalgo, Pachuca, Mexico

List of Figures

Fig. 1.1 Representation of complex systems states and phase transitions according to different schools and approaches. *Source* Authors' elaboration 10
Fig. 2.1 Relationship between market view(s) and market shaping strategy to be shared by collaborators for market shaping. *Source* Authors' elaboration 31
Fig. 2.2 Synthesis of the methodology and contributions of the manuscript. *Source* Authors' elaboration 32
Fig. 2.3 Collective market shaping. The upward and downward arrows represent the interactions between actors, resources, and emergent properties. *Source* Authors' elaboration based on the fourth order of emergence of Vargo et al. (2023) 37
Fig. 3.1 The duality of complexity in service ecosystems dynamics elaboration. *Source* The authors' 58
Fig. 3.2 Service ecosystem states and service ecosystems evolution according to an institutional view. Elaborated from Sarno et al. 2025a) 59
Fig. 3.3 Frictions—dilemmas, trade-off and paradoxical tensions—in complexity elaboration. *Source* The authors' 61
Fig. 3.4 The complexity states of service ecosystems, the related frictions and the possible service ecosystem dynamics towards evolution elaboration. *Source* The authors' 65
Fig. 4.1 The role of transdisciplinary systemic-evolutionary perspective. *Source* Author's elaboration 78
Fig. 4.2 Describing socio-economic systems in terms of complexity. *Source* Author's elaboration 89
Fig. 5.1 Nested levels of interactions for creating long-term and multi-level impacts in the municipalities 99
Fig. 5.2 Framework for impacts co-created in service ecosystems. *Source* Authors' elaboration 104

Fig. 6.1 Engagement influences actor disposition and resource integration in open platforms. *Source* The authors 117
Fig. 6.2 Crowdfunding ecosystem strategies for innovation emergence. *Source* The authors 119
Fig. 6.3 The resource-engagement innovation strategy matrix. *Source* The authors 121
Fig. 6.4 Rewards-based Crowdfunding Platform ecosystem. *Source* The authors 123
Fig. 7.1 The value depletion circle. *Source* Adapted from Díaz-Méndez and Saren (2019) 137
Fig. 7.2 VDC in HE. *Source* Own elaboration 138
Fig. 8.1 Multichannel versus Omnichannel customer journey. *Source* authors' elaboration 154
Fig. 8.2 Benefits from using chat assistants. *Source* Capgemini Research Institute Report 2019 155
Fig. 8.3 Customer satisfaction. Authors' elaboration from Capgemini Research Institute Report 2019. NPS: Net Promoter Score® (NPS) 155
Fig. 8.4 Conceptual framework. *Source* authors' elaboration 157
Fig. 8.5 The EA matrix. *Source* authors' elaboration 159
Fig. 9.1 NFT-based service ecosystems. *Source* Authors' elaboration 180
Fig. 10.1 Study's data structure. *Source* Authors' elaboration 196
Fig. 11.1 The paradigm of proprietary capitalism. *Source* Authors' elaboration 214
Fig. 11.2 The paradigm of possession. *Source* Authors' elaboration 216
Fig. 11.3 The paradigm of participation. *Source* Authors' elaboration 219
Fig. 11.4 'World 1' and 'World 2'. *Source* Authors' elaboration 221
Fig. 11.5 Global wealth distribution. *Source* Credit suisse 2021 222
Fig. 11.6 Emerging technology trends. *Source* Hackermoon.com 223
Fig. 11.7 The draining of resources. *Source* Authors' elaboration 227
Fig. 11.8 Implementing Re_cognition Income. *Source* Authors' elaboration 228
Fig. 11.9 Re_cognition towards 'World 1'. *Source* Authors' elaboration 229

List of Tables

Table 1.1	Approaches to phase transitions in complexity science	5
Table 1.2	Reconciliation of terminology adopted in complexity science approaches to phase transitions and advancements of S-D logic	11
Table 1.3	Examples of future research questions deriving from the study	19
Table 2.1	Extent of change the market system, changed "element," and related characteristics	40
Table 2.2	Typologies of systems' change and similar concepts	41
Table 2.3	Implications and future research related to the market views identification (detailing)	46
Table 2.4	Implications and future research related to the typology of market changes (differentiating)	48
Table 5.1	Theoretical underpinnings for steering STs in municipalities	102
Table 5.2	Summary of the key research activities in the design science process of this study	105
Table 7.1	Student as a co-creator of value	141
Table 8.1	Main constructs of EI by Goleman (1995)	152
Table 9.1	Sample	174
Table 9.2	Further research	181
Table 10.1	Interviewee titles and category	194

Chapter 1
Exploring Radical Market Changes as Phase Transitions of Service Ecosystems: Insights from Complexity Science

Debora Sarno, Jaqueline Pels, Francesco Polese, David Sörhammar, Bård Tronvoll, and Stephen Louis Vargo

Abstract The multiple current transformations can determine complex radical market changes. Many market actors feel an increasing need to make sense of market dynamics, which bring opportunities but also threats. We leverage on the Service-Dominant logic and deepen its roots in complexity science to delineate radical market changes (i.e., phase transitions of service ecosystems). In particular, within complexity science literature, we found that phase transitions are approached in three diverse ways: shifts to a new complex order, shifts to a region of complexity where chaos and order achieve a balance, and shifts to escape chaos. We show that those three approaches can be integrated to give a full explanation of the dynamics of the service ecosystems. Then, we reconcile the approaches with the S-D logic narrative to delineate the complexity science insights into the phase transitions of service ecosystems. In doing so, we highlight some interesting concepts to further explore within S-D logic: institutional instability and service ecosystem viability

---

D. Sarno (✉)
University of Naples Parthenope, Napoli, Italy
e-mail: debora.sarno@uniparthenope.it

J. Pels
Universidad Torcuato Di Tella, Buenos Aires, Argentina
e-mail: jpels@utdt.edu

F. Polese
University of Salerno, Fisciano, Italy
e-mail: fpolese@unisa.it

D. Sörhammar · B. Tronvoll
Inland Norway University of Applied Sciences, Rena, Norway
e-mail: david.sorhammar@inn.no

B. Tronvoll
e-mail: bard.tronvoll@inn.no

S. L. Vargo
University of Oklahoma, Norman, US
e-mail: sv@ou.edu

D. Sarno et al. (eds.), *Complexity and Emergence in Market Ecosystems*, New Economic Windows, https://doi.org/10.1007/978-3-031-81942-1_1

near the edge of chaos, and institutionalization due to self-organized criticality. The implications connect the conceptual findings to the practice of market actors facing transformations.

Keywords Value co-creation · Service ecosystems · Complex adaptive systems · Digital transformation · Complexity science

Introduction

Markets are not stable backdrops to business but may be seen as complex adaptive systems (Holland 2014) able to produce novelty and surprises (Markose 2005; see also Giesler and Fischer 2017), especially when introducing or applying new technologies, such as blockchain and cryptocurrencies (Allen et al. 2022), generative AI (Fui-Hoon Nah et al. 2023), and others. Thus, markets can be conceived as made up of a variety of networked actors (suppliers, customers, third-party organizations, authorities, etc.) integrating resources among which technologies (Lusch and Vargo 2014) for co-creating value (Blaschke et al. 2019). In this view, the multiple current digital transformations (Vial 2019) do not only occur at the company level but may determine complex radical market changes with environmental, societal, and institutional implications (Kaartemo and Nyström 2021; Kraus et al. 2021), as in the case of the current transition towards renewable energy sources around the world (Sarno and Siano 2022). Thus, market actors feel an increasing need to make sense of market dynamics (Nenonen et al. 2014), especially the large-scale step changes, since change can bring opportunities (such as operational efficiency or increased market share) but also threats (as loss of competitiveness or market dissolution). Moreover, it should not be neglected that, although market changes are complex, they can be somehow triggered (Li 2020), facilitated, conditioned, or prevented and, in some cases, computed and forecasted, and in general, must also be understood if the "simple" adaptation is pursued (Kraus et al. 2021).

Service-dominant logic (S-D logic) (Vargo and Lusch 2004, 2008, 2016), acknowledged as a systems-based general theory of markets (Akaka et al. 2021), can contribute to understanding radical market changes by observing them through the lenses of the service ecosystem (Vargo and Lusch 2016). The concepts of phase transitions have recently enriched S-D logic to describe radical system changes (Polese et al. 2021) entangled with institutional change (Kijima et al. 2016; Polese et al. 2021; Pels 2020; Pels and Mele 2022). This means to make sense of radical market changes not just for the advantage of a focal market actor (Peters et al. 2020; Cantwell et al. 2010) but to observe the overall evolution of the system. In other words, we aim to leverage S-D logic and deepen its roots in complexity science (that deals with complex adaptive systems as markets are) to better delineate radical market changes (i.e., phase transitions of service ecosystems). In particular, synthesizing previous definitions and building on the insights from complexity science, we define phase transitions of service ecosystems as a *shift to a new service ecosystem order near the edge of chaos.*

Thus, we first introduce S-D logic and phase transitions of service ecosystems (Sect. 1.2). Then, we present and integrate diverse (Phillips and Ritala 2019) complexity science schools of thought and transition approaches (Sect. 1.3). This allows us to reconcile such diverse perspectives through S-D logic and delineate the insights of complexity science into service ecosystem phase transitions (Sect. 1.4). The presentation of the theoretical contributions (Sect. 1.5) and the future research questions (Sect. 1.6) follow. Finally, in Sect. 1.7, the implications connect the conceptual findings of the chapter to the practice of market actors facing digital transformations.

S-D Logic and Phase Transitions of Service Ecosystems

S-D logic is an acknowledged theoretical framework indigenous to the marketing field. Lusch and Vargo (2014: 24) define service ecosystems as a "relatively self-contained, self-adjusting system of resource-integrating actors connected by shared institutional arrangements and mutual value creation through service exchange." Service ecosystems have become foundational to marketing studies since they allow underscoring the systemic nature of value co-creation, which is a multi-actors process enabled (and constrained) by institutional arrangements, such as shared rules, norms, symbols, languages, beliefs, etc. (Vargo and Lusch 2016). Moreover, adopting a complex adaptive systemic approach allows analyzing market dynamics, co-evolution, and co-viability with other nested and overlapping ecosystems. S-D logic adopts a multi-level perspective, which considers service ecosystems multi-layered and that different viewpoints can be used to analyze ongoing phenomena (Vargo and Lusch 2016).

Within S-D logic, work on change (Kleinaltenkamp et al. 2018; Tuominen et al. 2020) and large-scale step changes, i.e., phase transitions (Polese et al. 2021), are mostly related to institutions and institutional arrangements. These studies are aligned with the extant studies that have looked at links between S-D logic and the role of institutions, institutional arrangements, and institutionalization (e.g., Pohlmann and Kaartemo 2017; Vargo and Lusch 2017; Brodie et al. 2019; Vargo et al. 2020). Near to our interest we find: (1) works associating the importance of agreed norms and standards to achieve new meanings (Peñaloza and Mish 2011); (2) studies linking innovation to the result of ongoing negotiations and recombination of overlapping and intersecting institutions (Akaka et al. 2021; Kaartemo et al. 2018) and institutional complexity (Siltaloppi et al. 2016); (3) research on institutional work seeking to understand how service ecosystems change (Nenonen et al. 2018; Lintula et al. 2020); and (4) articles on the emergence of institutions and service ecosystems adopting agent-based modeling (ABM) (Fujita et al. 2018).

Specifically, potential sources of institutional change are: external environmental disturbances to the service ecosystem as megatrends not aligned with current institutional arrangements creating tensions (Kleinaltenkamp et al. 2018); actors' resource integration bringing the emergence of new service ecosystem properties (Peters 2016; Polese et al. 2020; Vargo et al. 2023a, b). Accordingly, Polese et al. (2021) define the phase transition process of a service ecosystem as a large-scale step change that occurs when external environmental disturbances and internal actors' resource integration dislodge the ecosystem from a state of stability into de-institutionalization and re-institutionalization, ending with the achievement of a new stable state. Such a new state is characterized by new institutional arrangements and value as a new organizing principle providing order and organization to the actors' interactions in the service ecosystem. Polese et al. (2021) were mainly based on literature from physics and the theory of synergetics, thus mainly focused on one of the approaches to phase transitions available in the complexity science literature. Moreover, considering complexity science and the collective nature of the phenomena, size has been recognized as central to achieving the diffusion of change (i.e., increasing returns to scale) (Vargo et al. 2020).

In short, within the S-D logic narrative, service ecosystem and phase transitions have been described as linked to the institutional change literature. However, we agree with Pohlmanna and Kaartemo (2017) signaling that institutions have become crucial to the development of S-D logic but need further development, as well as Vargo and Lusch's (2017: 55) statement that the role of institutions and institutional arrangements, in the future, "is to become considerably more evident in S-D logic." In other words, within the literature on service ecosystems, a unified understanding and description of the mechanisms of how phase transitions come about and the role played by institutional change is still missing.

Integrating Complexity Science Approaches to Phase Transitions

Complexity science and general systems theory (von Bertalanffy 1972) seek to provide a general understanding of (complex) systems behavior. Insights from physics (Prigogine and Stengers 1984) and biology (Kauffman 1969) have been applied to diverse fields, such as economics (Arthur 2021) and social contexts (Mathews et al. 1999).

Among the diverse types of systems, we focus on literature related to complex and complex adaptive systems (CAS) because they are more appropriate to explain complex social phenomena such as phase transitions. The main features of *complex systems* are (for details, see Cillers 1998): (1) they aggregate heterogeneous components—actors and resources–that interact, giving rise to emergent properties of the whole (so that the properties of the whole cannot be predicted by nor reduced to the properties of the components); (2) they incorporate amplifying (positive) and

Table 1.1 Approaches to phase transitions in complexity science

	Schools dealing with complexity science			
	European school		American school	
Approaches to the concept of phase transition	Focus of the phase transition	System-environment processes	Intra-system processes	
	Definition of the phase transition	Shift to a new complex order out of chaos	Shifts to a region of complexity where chaos and order achieve balance	Shifts to a new complex order to escape chaos

Source Authors' elaboration

balancing (negative) feedback loops—chains of causality which are deviation amplifying and deviation counteracting—that foster or prevent systems change (Capra 1996); and (3) they have a non-linear behavior (so that small causes can have disproportionate effects—a famous example is the Lorentz's butterfly effect, according to which the small butterfly flapping its wings could, hypothetically, cause a typhoon, 1963). *Complex Adaptive Systems* add to complex systems adaptive capacities such as "self-generation, self-organization, decentralized control, memory, evolutionary and concurrent persistence and change (resilience), and anticipatory capacities" (Preiser et al. 2018 p. 5). In this chapter, we deepen these elements and show how they contribute to unfolding phase transitions in service ecosystems.

As mentioned, a phase transition is a significant system shift (Howe and Lewis 2005; Dooley 1997; Jantsch 1980). However, the justifications for such a shift seem profoundly heterogeneous and contradictory. To make order, we have first revised the main classifications of complexity science studies. In particular, Maguire et al. (2006) identified two Schools of thought/origin: the European one (e.g., Prigogine, Haken, etc.) with a focus on *system-environment processes* and dealing with "order out of chaos" system condition; and the North American one (e.g., Holling, Kauffman, and Arthur from the Santa Fe Institute) with a focus on *intra-system processes* and dealing with the "edge of chaos" system condition. Furthermore, focusing on the origin of the shift of the transition, a literature review allowed us to identify three basic approaches: (i) shifts to a new complex order; (ii) shifts to a region of complexity where chaos and order achieve a balance; and (iii) shifts to escape chaos. In the following subsections, we discuss each of the three approaches (see Table 1.1).

Phase Transitions as a New Complex Order Out of Chaos

The overall focus of the European School was on the interactions of complex systems–intended as "united wholes"–with their environment and the related tensions that would trigger a phase transition (Maguire et al. 2006).

This stream of research originated in Europe with studies on complex physical systems. It analyzed systems' ability to reconfigure into new ordered structures when undergoing energy forces brought to a certain critical level. In particular, Prigogine and Stengers (1984) describe how complex systems far from equilibrium can maintain their stability by producing energy, exchanging energy with the environment, and dissipating energy. Moreover, too much energy in a system that cannot be absorbed or dissipated by it can bring high instabilities and chaos. However, when a tipping/bifurcation point is achieved, new structures and forms of order may emerge spontaneously (Glansdorff and Prigogine 1971), thus causing both the old and the new systems to branch off into divergent paths. In these situations, the fundamental structures of those systems, called dissipative structures, evolve. In other words, order emerges "out of chaos" through self-organization (Prigogine and Stengers 1984). Thus, self-organization sustains the creation of emergent novel properties and modes of behavior that cannot be reduced to single elements (Fuchs 2003) and is due to the interactions of the components of the system in the process of development, learning, and evolution (Capra 1996), fostered and constrained by amplifying and balancing feedback loops (Lichtenstein and Plowman 2009; MacIntosh and MacLean 1999). Other studies in this area are related to the theory of synergetics, dealing with order, control parameters, and bifurcation points (Haken 1983) (bifurcation theory, instead, focuses on the discontinuities and instabilities driving them–for an overview, see Eidelson 1997).

The phase transition process according to this approach can be summarized as follows:

High instabilities–due to positive and negative feedback loops involving emergent properties–bring the system to a bifurcation point (disruption of equilibrium). Here, chaos is avoided thanks to self-organization, and a new order emerges.

Phase Transitions as a Region of Complexity Where Chaos and Order Achieve Balance

The American School studies on complexity science emphasize "intra-system processes resulting in emergent complexity and, in particular, the co-evolution of parts to a state [...] beyond which order disappears, and disorder overwhelms the system" (Maguire et al. 2006, p. 169).

Based on the developments in computer science and drawing from life science, the American School focuses on heterogeneous agents and their co-evolution. In particular, agents are considered search for fitness (that is, agent's goodness) and thus able to adapt to the environment by changing their schemas (i.e., rules for interpretation and actor or, in other words, rules for fitness rewards). Changes at the micro-level can also depend on actions to adapt observations to existing schemas or changes to schemas due to adaptation to observations or combinations with other schemas. Furthermore, the heterogeneity and reactivity of agents (Rogers et al. 2005) can allow the diffusion of the new schemas and aggregations around them at the macro level (Dooley 1997). In this school, computational models are being developed to study interactions resulting in collective phenomena, trying to predict behaviors.

The American School pays particular attention to the notion of a system's "edge of chaos" state. The "edge of chaos" refers to the precarious balance of order and chaos (Holbrook 2003). Waldrop (1992: 12) clearly explains: "the edge of chaos is where new ideas and innovative genotypes are forever nibbling away at the edges of the status quo, and where even the most entrenched old guard will eventually be overthrown. ... The edge of chaos is the constantly shifting battle zone between stagnation and anarchy, the one place where a complex system can be spontaneous, adaptive, and alive". At the edge of chaos, the disorder can create opportunities for learning and adaptability and preserves viability across changing conditions (Smith and Gemmill 1991); thus, this state is considered the only one where shifts can be viable. An interesting associated concept is "self-organized criticality." According to it, the "size and frequency of restructuring events ... [of a phase transition] are related by an inverse power law" (Maguire et al. 2006, p. 167), which describes how propagation occurs at the edge of chaos. However, to avoid too much novelty that could overwhelm a system, some scholars believe that complex systems should be kept "near" and not "at" it (Anderson 1999). Similarly, many people experience their most productive moments near the temporal edge of the chaos of a deadline and learn how to make constructive use of an upcoming time limit (Pascale et al. 2001). In other words, the co-evolution toward the edge of chaos can bring the selective advantage of agents due to slight increases in fitness levels. At the same time, too big changes in fitness values (which can occur in chaotic environments) may not last long and result in lower fitness levels (Kauffman 1993).

The phase transition process of this approach can be summarized as follows:

High change (in the agent's schema) results from the interplay between local (micro level) and global (macro level) optimization of agents' fitness. Such change brings the system to the edge of chaos, where there is an increase in the fitness levels of the system, and chaos and order achieve balance.

Phase Transitions as a New Complex Order to Escape Chaos

Within the American School, some researchers considered chaos not as a state that needed to be balanced with order (as in the previous approach) but as a situation where the system might reach dissolution. Thus, they considered transition processes that allow complex adaptive systems to escape chaos and dissolution.

According to this approach, the system's dynamics are mathematically explained through power laws, showing that some metrics of the systems can increase more than proportionally. Here, the need for change depends on the co-evolution of parts composing the complex system that pushed the system towards chaos.

Moreover, the American School also developed social-ecological systems studies, introducing the concept of adaptation in complex systems (Gunderson and Holling

2002). Complex adaptive systems (CAS) draw on complexity science constructs to focus on human interactions and adaptation with/within nature (Colding and Barthel 2019). Though there is no unified CAS perspective (Preiser et al. 2018), generally, phase transitions have been termed "transformations" and defined as the "capacity to create a fundamentally new system when ecological, economic, or social structures make the existing system untenable" (Walker et al. 2004, p. 1). In particular, Gunderson and Holling (2002) dealt with the adaptive pattern of ecosystems, explaining it as cyclical and made up of phases of increasing growth of the system that cannot be sustained and followed by renewal where some mutations, variations, or novelties previously occurred can be the source of a "new order" and result in a phase transition. Furthermore, they introduced hierarchical ontological levels within the ecosystem to explain change at some levels and cascade events to others.

Geoffrey West's study on scaling has contributed to understanding CAS behavior and related transitions (West 2017). 'Scaling' refers to how a system responds when its size changes, with a particular focus on cities. West observed that when a city's population grows, many of its socio-economic metrics (both the positive ones, as the GDP, and the negative ones, as crime) increase more than proportionally. West labeled this characteristic superlinear scaling and argues that "this kind of growth behavior is clearly unsustainable because it requires an unlimited, ever-increasing, and eventually infinite supply of energy and resources at some finite time in the future to maintain it. Left unchecked [...], it triggers a transition to a phase that leads to stagnation and eventual collapse [...] Major innovations can therefore be viewed as mechanisms for ensuring [...] a transition from one phase of the system to another having very different characteristics [...] circumnavigating the potentially disastrous discontinuity." In other words, paradigm-shifting innovations are presented as a way to "reset the clock before potential collapse occurs" and "ensure open-ended growth" (West 2017, p. 416).

The phase transition process of this approach can be summarized as:

The increasing growth of variables characterizing agents in the system due to power laws makes the system unsustainable, directed toward chaos and collapse. A phase transition may occur driven by mutations, variations or novelties, and influences among levels.

Integrating Complexity Science Approaches on Phase Transitions

We draw on definitions of phase transitions from physics to integrate the different approaches to phase transitions provided in the previous subsections (McKelvey 2001; Maguire et al. 2006). McKelvey (2001), drawing on Mainzer (1997), identified two critical points (intended as values of energy of specific control parameters) a complex system can experience and which can change complex systems' behavior. The first critical point (or first-order phase transition) signals a transition from one state of the system to another, where the import of a certain energy from the environment is needed for the transition to occur. An example in physics is the liquid–solid transition when water becomes ice under certain temperature and pressure conditions. If we consider water as the chaotic phase and ice as the ordered one, "the molecules are forced to make an either-or choice between order and chaos" (Waldrop 1992). The second critical point (or second-order phase transition) is characterized by a continuous exchange of energy, which could also be infinite. An example is the transition from paramagnetic to ferromagnetic states or, in other words, when the material becomes magnetic due to a field force and stays magnetic, also when the field force is taken away. These transitions "are much less abrupt, largely because the molecules in such a system do not have to make that either-or choice. They combine chaos and order." (Waldrop 1992). The two critical points demark different kinds of phase transitions and define the upper and lower bounds of a region of emergent complexity (McKelvey 2001). Thus, thanks to the distinction between the two types of phase transitions in physics, we can see that the complexity science approaches to phase transitions refer to diverse things. Indeed, while the European School identified the phase transition (with the emergence of a complex order) as the first critical point, the American School dealt with the phase transition (with self-organized criticality) as the second critical point.

Thus, we argue that a full explanation of the dynamics of complex systems requires understating both Schools' and their three approaches. We developed Fig. 1.1 to help visualize this integration. It shows a spiral that highlights a complex (adaptive) system evolution. This spiral operates within the region of emergent complexity delimitated by order and chaos. Within this space, the systems experience change from the environment (first-order phase transition) and within (second-order phase transition). The spiral shows a sequence of phase transitions, where a new order replaces a previous (unsustainable, according to the second approach) one that was becoming chaotic to escape chaos. In contrast, the balance between chaos and order is sometimes sought to foster creativity and change. We notice that the first approach might seem similar to the third one, but the change that the systems experience in the latter is not sharp nor definitive and depends on the co-evolution of parts instead of inter-system changes, which might be cases of "less complex" systems.

Having integrated the main approaches to phase transitions in complexity science, we move to delineate how these insights relate to and contribute to our understanding of transitions in service ecosystems, particularly the role of institutions and institutional arrangements in service ecosystems.

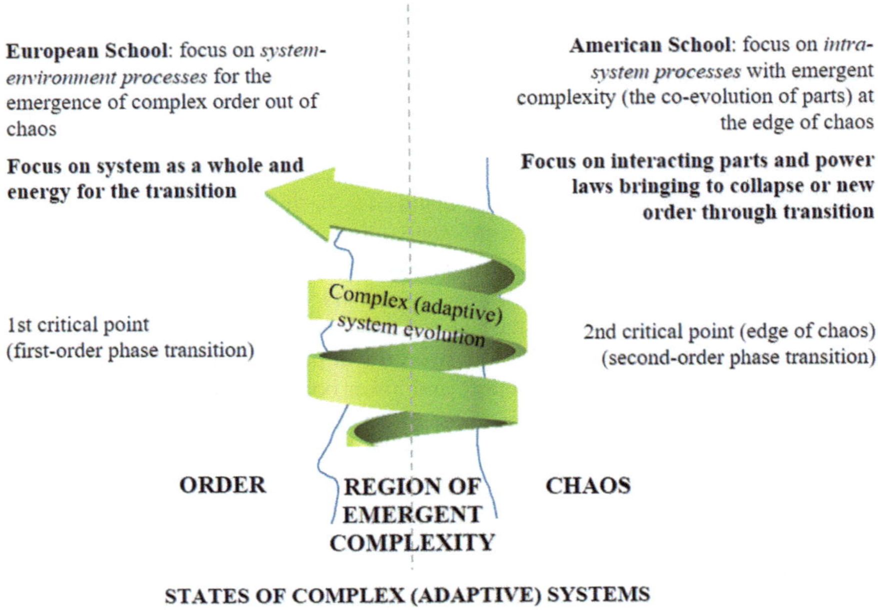

Fig. 1.1 Representation of complex systems states and phase transitions according to different schools and approaches. *Source* Authors' elaboration

Delineating Complexity Science Insights into Phase Transitions of Service Ecosystems

Having shown, in Sect. 1.2, the S-D logic state-of-the-art on phase transitions and institutional change and having integrated the three complexity science approaches to phase transitions of complex systems in Sect. 1.3, we now provide an S-D logic based reconciliation of the terms used in those approaches (see Table 1.2). Read with S-D logic lenses, phase transitions are affected by inter (otherwise named external disturbances) and intra-ecosystem processes (or internal actors' interactions) at macro, meso, and micro levels. As a result, S-D logic can leverage the two complexity science schools and accommodate them. Moreover, a service ecosystem may go through all the dynamics and evolution described in Fig. 1.1.

In Table 1.2, the first column shows the classical S-D logic narrative ("the organizing framework"), and the next columns represent the three complexity science insights ("the informing frameworks"), while the last column synthesizes the advancement of S-D logic due to incorporating the insights from complexity science. In the following subsections, we discuss and compare the narrative of the diverse approaches to phase transitions in complexity science with the S-D logic narrative moving over the rows of Table 1.2. Synthesizing previous definitions of phase transitions and building on the insights from complexity science, we define

Table 1.2 Reconciliation of terminology adopted in complexity science approaches to phase transitions and advancements of S-D logic

S-D logic narrative	Complexity science approaches to phase transitions (applications to social studies are mentioned in brackets)			S-D logic narrative incorporating phase transitions
	Shift to a new complex order out of chaos (emergence of new organizing principles)	Shifts to a region of complexity where chaos and order achieve balance	Shifts to a new complex order to escape chaos	Shift to a new service ecosystem order near the edge of chaos
Actors	Elements	Agents—characterized by heterogeneity and reactivity	Agents	Actors
Resources	Energy (information and other resources)	Resources	Resources	Resources
Emergence	Emergence	Emergence	Emergence (mutations, variations, and novelties)	Emergence
Value	(organizing principle)	Fitness (satisfaction in the system)	-	Value
Institutional arrangements	Dissipative structures (archetypes, mental models)	Rules for fitness rewards (schema; social norms)	Structures of the agents' networks underlying wealth creation	Institutional arrangements
Institutional complexity	Instability, high increase of energy (activities/events occurring outside the norm, lack of organizational coherence)	Lack of fitness of agents	Systems close to chaos, i.e., lack of resources to follow power laws	Institutional instability (edge of chaos)

(continued)

Table 1.2 (continued)

S-D logic narrative	Complexity science approaches to phase transitions (applications to social studies are mentioned in brackets)			S-D logic narrative incorporating phase transitions
	Shift to a new complex order out of chaos (emergence of new organizing principles)	Shifts to a region of complexity where chaos and order achieve balance	Shifts to a new complex order to escape chaos	Shift to a new service ecosystem order near the edge of chaos
Institutionalization -dealing with self-adaptation and self-adjustment	- Amplifying (also called reinforcing or positive) feedback loops amplify the reproduction/diffusion of emergent properties - Balancing (also called stabilizing or negative) feedback loops anchor the change by slowing the amplification of the diffusion of emergent properties - Self-organization	- Feedback between local and global optimization of agents' fitness - Feedbacks act at each level and among levels - Self-organization - Self-organized criticality	Power laws relating systems metrics among ecosystem levels	Institutionalization—dealing with self-adaptation, self-adjustment, and self-organized criticality
Service ecosystem viability	Far from equilibrium conditions, dynamic stability/equilibrium, order	Better fit at the edge of chaos	Avoidance of chaos	Service ecosystem viability, a system near the edge of chaos

Source Authors' elaboration

phase transitions of service ecosystems as "*a shift to a new service ecosystem order near to the edge of chaos.*"

Actors, Resources, and Emergence The concepts of actors (in complexity science called elements, agents, or individuals), resources, and emergence are aligned with those described by the S-D logic (Vargo et al. 2023a, b); thus, they will not be discussed in detail.

Value (individual or collective) is a core organizing principle within service ecosystems (Meynhardt et al. 2016). This view is aligned with Polese et al. (2021) concept of the emergence of a new order (which draws from the concept of the order parameter of a phase transition). Also, as discussed in the second complexity science approach, the S-D logic concept of individual/intersubjective phenomenological determination is represented by the notion of local/global agent fitness, which, in organizational terms, means satisfaction in the system (Dooley 1997). Thus, the actor satisfaction in a system determines if it will try to maintain or change its status quo through changes in its practices and other institutional arrangements (for value co-creation) with other actors.

Institutional arrangements Structures—ruling and ruled by actors' interactions (Giddens 1984)—are recognizable in each of the three complexity sciences approaches and are compliant with the S-D logic narrative on institutional arrangements. Moreover, applications in organizational and management literature mentioned archetypes, schema (MacIntosh and MacLean 1999), mental models (Senge 1990), and social norms (Rogers 2003), directly referred to institutional theory (e.g., Giddens). When linked to the concept of fitness (discussed in the previous paragraph), institutional arrangements are considered rules for fitness rewards for actors (Dooley 1997). It is essential to note that institutional arrangements are not guarantees of optimal solutions but are often reasonably fitting to the surrounding situations (Fujita et al. 2019). Schema may adapt to a local optimum (Kauffman 1995) and become deeply ingrained and difficult to alter, thus representing maladaptation over the long run (Gell-Mann 1994). The "QWERTY" keyboard is an often-quoted example of such maladaptation (Arthur 1994).

Institutional complexity The emergence and institutionalization of new service ecosystem properties, such as new actors, resources, value, and institutional arrangements (Polese et al. 2021), can determine institutional complexity, i.e., a multiplicity of institutional arrangements with conflicting prescriptions for action that actors experience (Siltaloppi et al. 2016). Institutional complexity has diverse consequences in the service ecosystem. The deriving tensions between institutional arrangements can reduce the influence of prevailing institutions–and may drive to de-institutionalization; simultaneously, the new properties can also make new institutional "toolkits" available to actors, which may drive to re-institutionalization.

Within the European School, institutional complexity is captured by the word "instability," where it is described as initiated by activities or events occurring "outside the norm" in comparison to the context (Lichtenstein and Plowman 2009), or by new interpretative schemas or mental models. Such emergence is concurrent with increased instabilities and the loss of organizational coherence (MacIntosh and MacLean 1999).

The second approach describes the double face of institutional complexity, which is identified as a lack of fitness of agents, leading towards a change in the schema of agents (Dooley 1997). Along the same lines, the maladaptation mentioned above could be non-beneficial to the actors to the point of acknowledging its negative impact. As a result, when institutional arrangements lose their reasonable fit, it is often an indicator that there will be a change. The lack of fitness can lead the current pattern of the system towards the edge of chaos. As Waldrop (1992) wrote, "complex systems have somehow acquired the ability to bring order and chaos into a special kind of balance. This balance point–often called the edge of chaos–is where the components of a system never quite lock into place, and yet never quite dissolve into turbulence, either. The edge of chaos is where life has enough stability to sustain itself and enough creativity to deserve the name of life." Thus, although institutional complexity may seem detrimental to value co-creation because value loses meaning and may confuse actors, it is also a source of renewal. Here, the importance of heterogeneous interactions to bring adoption, mutation, or change is highlighted. Rogers et al. (2005) focused on two properties of CAS: variety/heterogeneity, already reported by Vargo et al. (2020), and reactivity. We highlight that both are related to the edge of chaos, a heterogeneous service ecosystem state where emergence is the likeliest. In such a state, the actors, in their variety, "can react more sensitively, increase their fitness, and change in a way that enhances chances for survival or forestalls threats of extinction." (Waldrop 1992, p. 6).

The consequent renewal is the objective of the third approach, where mutations, variations, and novelties can be the source of a "new order" in the growing phase of an ecosystem (Gunderson and Holling 2002). Here, the institutional complexity is due to the untenability of the current systems with its current institutional arrangements. In other words, the finite time singularity (according to superlinear scaling, it can be identified as a precise time in which the growth of the socio-economic parameters becomes infinite) described by West (2017) will happen, and the system must be renovated to survive.

We believe that the edge of chaos concept significantly contributes to understanding phase transitions within S-D logic as it captures the importance of institutional complexity and its role in making a service ecosystem viable. However, too much institutional complexity can also cause a system to collapse. Thus, we agree with the third approach rationale that the system must be near the edge of chaos but not at the edge of chaos. In other words, the system must constantly evolve, as in the spiral of Fig. 1.1, to stabilize/institutionalize for a while, taking advantage of the changes before changing again. This is why we argue that the expression 'institutional complexity,' though correct, does not capture the potential misalignments. As a result, we suggest adopting the label Institutional Instability (Edge of Chaos) in the last column of Table 1.2.

Institutionalization The abovementioned dynamics could result in the self-organization of the system. S-D logic has explained that service ecosystems are self-adaptive and self-adjusting systems responding to internal and external variations using positive and negative feedback loops (Vargo et al. 2023a, b). In other words, some ephemeral emerging properties can become institutionalized as proto-institutions (Kleinaltenkamp et al. 2018) or fully-fledged institutions. Though with a diverse focus, the interplay between feedback loops and self-organization is discussed in all three approaches to phase transitions. Here, we want to highlight the differences among the three approaches within complexity science and how each can enrich S-D logic's understanding of a phase transition.

According to the European approach, the feedback loops are systemic forces that can depend on the system-environment relationships and then connect service ecosystem components because they are embedded in institutional arrangements. Both amplifying (positive or reinforcing) feedback loops and balancing (negative or stabilizing) feedback loops act on the overall system. The former (amplifying) fosters the diffusion of emergent properties and results in deinstitutionalizing pressures (Oliver 1992; Greenwood and Hinings 1996) due to questioning taken-for-granted assumptions. Consequently, actors' self-organization is needed to emerge a new order. The latter (stabilizing) highlights the importance of anchoring and institutionalizing the change by slowing the non-linear processes led by the previous amplification of the diffusion of emergent properties (Lichtenstein and Plowman 2009).

Moreover, the work of MacIntosh and MacLean (1999) does not only emphasize that feedback loops foster change but also that there are institutional tensions with the feedback loops already in place, which determine the status quo. Furthermore, the de-institutionalizing/re-institutionalizing tensions are evident in external pressures, internal politics, and random couplings that combine as institutional arrangements to establish themselves in the face of resistance from defensive routines (Argyris 1990).

In the second approach, the institutionalization is driven by feedback between local changes and global optimization of agents' fitness. Thus, the second approach highlights the system's dynamics at different aggregation levels. Rogers et al. (2005) show that a global change can happen when local changes are significant (*critical mass inflection point,* the central point in the S-shaped diffusion curve), a concept captured by self-organized criticality. A widespread example to explain it is the sandpile: when sand is added on the top of a sandpile, the slope of the pile grows until a critical angle is reached, then such slope remains constant while the sand flows along it in the shape of avalanches of different size and duration, following power laws (Bak et al. 1991). Thus, the inflection point where the rate of new adopters changes (and increases significantly) is a phase transition, a situation in which, at the edge of chaos, self-organized criticality manifests through avalanches, which are explained as mutations and changes through which actors adapt (Rogers et al. 2005). At this point, the system processes new information and overcomes uncertainty. Furthermore, feedback can act within one level (for example, individuals influence others) or between levels, as feedback from the macro-level to individuals.

Finally, in the third approach, although there is attention to the co-evolution dynamics of actors that drive the change, the feedback loops, on average, are highlighted at the system level. The reason is that they are embedded in the power laws that relate systems metrics of all actors, such as the average growth of the system over time.

Again, the three approaches complement each other because they draw diverse causal forces (amplifying and balancing) connecting actors from the outside, within, and between different levels of aggregation and driving the institutionalization of emerging properties. Thus, the three approaches allow us to enrich Vargo et al.'s. (2023) generic discussion of the dynamics between emergence and institutionalization, which resented a causal interplay between parts and whole. Furthermore, the introduction of self-organized criticality alongside self-organization and self-adaptation enables the depiction of the systemic behavior that pushes a wide set of actors to change as 'avalanches of a sand-pile' when the institutionalization of new properties has become so pervasive to reduce uncertainty about the future and hesitations for actions. Close to the edge of chaos, such a tipping point is the state of the highest institutional instability.

Service ecosystem viability From the first approach, we have learned that the new order achieved through a transition in complex systems means that systems are not "static" but, as Capra (1996) recognized, it indicates the presence of order or organization while the systems need to dynamically change (dynamic stability) to stay viable.

The second approach highlights that, at the end of the phase transition, the lack of fitness, which was the engine of the transition, is overcome. Overall, there is an increase in the fitness level of the system. Each agent's change (to improve their fitness) could have changed the fitness landscape of the other agents in their network of relationships (Rogers et al. 2005). Such "global optimization enhances the CAS coherence as a system" (Dooley 1997: 87). In other words, the system coalesces since individuals have risen to the group threshold of fitness and adaptation (development into a fitter large-scale system). From this moment on, in innovation diffusion, there are enough adopters for further diffusion to become self-sustaining (Rogers 2003). However, collapse could occur if actors are inhibited in their ability to adapt interdependently (Rogers et al. 2005) and reciprocally influence fitness landscapes. The concept of fitness and the interplay between actors' and global fitness in the co-evolutions of actors above is one of the most interesting insights from the second approach to phase transitions. Vargo and Lusch (2017: 56) mentioned it referring to inclusive fitness as "likely to be particularly useful in informing S-D logic." We agree that it can "contribute to the discussion of value, especially as conceptualized in terms of change in the viability of a system." The value (fitness) an actor perceives depends on the actor's context (fitness landscape), which must be related to institutional arrangements, the actor's network, and resources. A fitness landscape can be graphically represented as a range of mountains the actor tries to climb (the tallest peak represents the site of the actor's potential maximum value). At the same time, they dynamically change due to actors' interactions. Furthermore, to reach their respective fitness peaks, co-evolving actors must adapt to one another (Eidelson 1997), giving birth to a service ecosystem. "As a result, their optima are no longer fixed and independent; the [actors] experience their shared environment as a landscape that constantly shifts and deform" (Eidelson 1997, p. 58). Although emergence, bounded rationality, and agency may cause different pathways that the optimal one to be followed, it is also true that the fitness landscape can offer a way to visualize actors' possible pathways and preferable interactions for value co-creation and increase the viability of the overall service ecosystem.

Finally, the edge of chaos is where the service ecosystem viability is the highest, but the untenability is the maximum. This is why, as already anticipated when dealing with institutional complexity, a service ecosystem should be kept near the edge of chaos. This is supported by the third approach, in which a transition is needed to avoid the edge of chaos and keep viability through change.

Theoretical Contribution

Central to marketing and management is theorizing on how actors—buyers, sellers, and other engaged stakeholders—can be involved in or adjust to changes in the system (intra-system processes) or to the system (system-environment processes). These changes can either be small and incremental or create a phase transition for the whole system. Incremental variations occur when the overarching institutional arrangements stay relatively intact. Here, research has highlighted ways for actors to adapt new insights into their existing solutions (incremental innovations) and become more agile (Teece et al. 2016) and flexible (Hatum and Pettigrew 2006), suggesting actors constantly sense changes in the market, seizing these insights into the organization and then transform their activities (Teece 2007). However, less guidance has been given on how actors should act during phase transitions.

The chapter extends recent conceptualizations of phase transition grounded in S-D logic and its service ecosystems perspective (Vargo et al. 2023a, b; Polese et al. 2021) by reconciling insights regarding phase transitions from three distinct complexity science perspectives. By integrating (MacInnis 2011) these three perspectives and delineating (MacInnis 2011) phase transitions in the S-D logic narrative, our findings try to strengthen the relationship between complexity science and S-D logic and emphasize the institutional dimension of the phase transition. In particular, at least three distinct contributions can be highlighted.

First, this study illuminates the distinct nature of three specific approaches to phase transitions in complexity science. By teasing out their inherent specificities, the European school is stated to focus on system-environment processes, whereas the American school, with unique features, focuses on intra-system processes. However, we go one step further and integrate these perspectives in Table 1.1 and Fig. 1.1. By focusing on the relationship with the environment, the European School links the emergence of complex order at the first critical point of a phase transition. Alternatively, the American School–stressing the intra-system process–relates phase transition to self-organized criticality at the second critical point. Both Table 1.1 and Fig. 1.1 allow us to visualize that the three approaches hold a dialectical and complementary relationship and that, combined, they allow a holistic understanding of the transition phenomenon.

Second, by reconciling the three complexity science approaches with the S-D logic narrative, we enrich S-D logic's understanding of phase transitions in service ecosystems (delineating contribution). Specifically, we highlight alignment with three core concepts of the S-D logic narrative (actors, resources, and emergence) and discuss the differences in the other concepts. The complexity science literature's main contributions are associated with the institutional-related insights, particularly through the introduction of the concepts of edge of chaos, which led to suggest substituting institutional instability to institutional complexity into the S-D logic narrative, and self-organized criticality, which allowed to propose that the S-D logic notions of self-organized criticality and self-adaptation can be complemented to it to explain institutionalization.

Finally, the reconciliation of the three complexity science approaches highlights the importance for service ecosystems researchers not to focus on finding ways for systems to be stable since stability, unfortunately, leads to the demise not only for individual actors but for the system itself. Instead, they allow drawing attention to the dynamic stability of service ecosystems near the edge of chaos, able to continuously renovate for viability without risking collapse.

Additionally, by discussing the extreme situation of phase transitions, this chapter enriches the set of concepts adopted in the S-D logic literature to discuss institutionalization and institutional arrangements, such as dissipative structures, rules for fitness rewards, power laws, local/global interplay, avoidance of chaos, feedback loops to diffuse change and stabilize the system. Finally, a phase transition of a service ecosystem that is not anchored by feedback loops regarding adjustments in institutional arrangements and the creation of new solutions for new problems might force the system beyond the edge of chaos into dissolution.

Future Research

Value co-creation, embedded in the service ecosystem, is a concept that emphasizes the importance of coordination and collaboration between various actors, creating a need for a shared understanding of how 'we do things' in practice (Taillard et al. 2016). However, turbulence may occur, for example, in response to intense social, economic, and political pressures or technological developments. A phase transition can occur, which requires the dismantling or restructuring current institutional arrangements and the emergence of new ones (de-institutionalization and re-institutionalization) (March and Olsen 1989; Peters 1999; Vargo, Wieland, and O'Brien 2023a, b). Future research could be oriented to further explore the processes of de- and re-institutionalization under the lenses of the three approaches to phase transitions shown in this study, as detailed in Table 1.3. The main elements are mentioned in the following.

The first approach involves studying how ecosystems can transition from chaos to a new order through self-organization and how institutional arrangements influence this process. For example, there is a need to investigate how the existence of formal

Table 1.3 Examples of future research questions deriving from the study

Complexity science approach to phase transition	Future research questions
Phase transitions as a new complex order out of chaos	Conditions under which service ecosystems move from a state of relative stability to a state of instability Dynamics of dissipative structures and how they emerge out of chaos through self-organization Role of feedback loops in amplifying and balancing emergent properties and promoting self-organization Factors that enable or inhibit the emergence of novel structures and behaviors Role of technology in fostering this type of transition
Phase transitions as a region of complexity where chaos and order achieve balance	Conditions under which the service ecosystem moves from a state of order to chaos or vice versa and how this affects its ability to adapt and learn Role of the "edge of chaos" in enabling complex adaptive behavior in various service ecosystems Role of feedback loops and information flows in promoting or inhibiting the emergence of complex adaptive behavior in various service ecosystems Dynamics of agent-based models, their co-evolution, and how this leads to emergent phenomena at the macro level Role of technology in fostering this type of transition
Phase transitions as a new complex order to escape chaos	Conditions under which institutional change occurs and how this affects the behavior of individuals and organizations within the ecosystem Role of institutions in enabling or constraining complex adaptive behavior in various service ecosystems Role of power laws, feedback loops, and information flows in promoting or inhibiting institutional change and innovation in different domains New theories or frameworks for understanding institutional change and innovation and designing interventions that promote these processes in different contexts Role of technology in fostering this type of transition

Source Authors' elaboration

and informal institutions can facilitate or impede the self-organizing process and how institutional change can affect the emergence of new structures and behaviors in the ecosystem. Additionally, it explores how feedback loops can influence the self-organizing process and how they can be amplified or balanced to facilitate the emergence and institutionalization of new structures.

The second approach covers how actors can achieve value co-creation while balancing order and chaos and how institutional arrangements influence this state. For example, researchers can investigate how institutional arrangements can facilitate or impede resource integration near the edge of chaos and how institutional change can affect the balance between order and chaos. Additionally, research can explore how achieving the edge of the chaos state can be influenced by the heterogeneity of actors and the diffusion of new rules for fitness rewards and feedback.

The third approach to phase transitions is related to a shift from chaos to a new order involving a continuous but large-scale step change process (as approach one, but less abrupt, with a focus on internal and environmental conditions and processes) needed to escape chaos and system dissolution. This approach particularly emphasizes the role of emergence as a driver of change; innovation is considered a key enabler with unpredictable outcomes. In service ecosystems, it can involve deepening the factors influencing institutional change, such as institutional entrepreneurs, power dynamics, and external shocks. Additionally, future research can explore the implications of institutional change for actors, such as changes in their identity, practices, and resource allocation. Besides, it can investigate the role of communication and sense-making in facilitating institutional change and how organizations can manage institutional change processes to minimize disruption and maximize opportunities for innovation.

The concepts of de- and re-institutionalization have important implications for further developing the service ecosystems perspective in S-D logic. Future research could explore the above areas to further the understanding of the role of institutions and institutional arrangements.

The (institutional) dynamics of the service ecosystem evolving through all of those phase transitions, as represented in Fig. 1.1, can be rewarding in identifying key drivers of viability and dissolution in diverse contexts.

Implications

Actors' characteristics to drive change Research has noticed the existence of *actors' abilities* to drive radical change actively. For instance, gaining a first-mover advantage (Lieberman and Montgomery 1988) or finding a blue ocean (Kim and Mauborgne 2004) can be prosperous for the actor and later for the system's viability. According to the European School of Complexity Science, it can be argued that the ability to regenerate, prosper, and improve under adverse situations, e.g., antifragility (Taleb 2012), can be achieved by embracing uncertainty. That is, amplifying feedback loops that are fostered by encouraging novelty (Lichtenstein and Plowman

2009). In comparison with minor system adjustments, in which actors could try to sustain and leverage institutional arrangements to predict the future–while focusing on dynamic capabilities (Teece 2007)–during phase transitions, courses of action are not predictable (Wiltbank et al. 2006) as the past no longer resembles the present. In these specific situations, actors' creativity, opportunity-seeking, and improvisation (Sarasvathy 2001; Weick 1998) are more likely to enable value co-creation. It can relate to creating radically new types of offerings and routines or changing the 'rules of the game' for all engaged actors in the service ecosystem. Actors operating in electronic markets should be particularly aware of these topics because new technologies significantly transform actors' abilities. For example, the metaverse has been considered an enabler of "creativity-guided co-creation" (Schöbel and Leimeister 2023). Furthermore, attention should be paid to "innovators" or, as Nicolò Machiavelli proposed in his book "The Prince", "the introductors of new orders". Those actors have as enemies all the supporters of the "old order" and as weak defenders all the potential beneficiaries of the "new order." Machiavelli concluded that not only the innovators need the occasion to propose something new, but also the virtue of seizing the opportunity and, finally, the resources to diffuse it.

Actors' resources are fundamental for a phase transition since resource integration for value co-creation can trigger the emergence of new systems properties (Peters 2016), which may trigger feedback loops (Vargo et al. 2023a, b), driving the transition. In the third approach to phase transitions, this is apparent thanks to the concept of self-organized criticality. Indeed, although power laws and feedback loops will foster the change, the initial differences among engaged actors are crucial to initiating a transition. Some actors have more resources to address the unbalance in the system by working towards a more coherent institutional arrangement than actors with fewer resources. In a system where all actors lack resources or are commonly unwilling to partake in harmonizing the institutional arrangements, the whole system may keep itself near the edge of chaos. This hints towards the critical importance of engaged and powerful actors in a system, especially according to the last approach. On the other hand, considering the first approach to phase transition, resources are needed to allow for the emergence of a new dissipative structure and to achieve a new equilibrium. For example, to explore and exploit the potential of Generative AI and transform markets, technical models (both general purpose and customized ones) must be integrated into service ecosystems (Banh and Strobel 2023), considering their specific characteristics regarding institutional arrangements and actors. In other words, only a context-aware Generative AI can be a powerful resource to support actors in their decisions.

Emergence (of new solutions) Then, *emergence* in search for mutations, variations, novelties, and creation of new solutions is fundamental for phase transitions. Indeed, as the service ecosystem goes from one phase to another, such as from the Iron Age to the Middle Ages, it creates the need for numerous innovations to solve problems that did not even exist in the previous (st)age. Stated differently, without radically new inventions to anchor the phase transition, the system might instead cross the edge of chaos and become chaotic. According to the American School, these multiple adjustments will most likely create numerous heterogeneous interactions in the system, some of which will become the source of a "new order" or bring the systems to further institutional complexity and collapse at the edge of chaos. The heterogeneity of actors, for example, has been one of the strategies identified for designing digital services for smart cities (Oschinsky et al. 2022). Most of the time, the emergence of new technological solutions "punctuates" existing market equilibria (Clemons in Alt 2022, leveraging the punctuated equilibria theory of Eldredge and Gould (1972) in evolutionary biology) and can be both beneficial and dangerous for market actors. However, as Clemons suggested, if you compare extreme cases that may happen in the future, you will not be precise enough to identify the new stable state of the service ecosystem. However, you may be accurate enough to drive your change while navigating the transition.

Institutionalization Later, for a phase transition, it is crucial to give rise to emergent properties and institutionalize new properties, working around numerous new and old institutional arrangements and resources to provide new solutions to new problems based on/thanks to feedback loops from the system. Indeed, these feedback loops enable the new properties either to be built on previous instantiation of institutional arrangements, as electric cars resemble gasoline cars down to specifics such as where to fill in the gas/electricity, or radical new inventions, as in the utilization of technologies the world did not even know about beforehand (e.g., Gen AI revolution). Drawing linkages between innovation and phase transition from a complexity science perspective, it can be stated that a phase transition for Rogers et al. (2005) occurs in the middle of the S-shaped curve of innovation, while the phase transition for West (2017) occurs between two paradigm-shift innovations or in other words before a finite time singularity is overcome thanks to the diffusion of one or more large scale step-change innovations. This can also bring to consideration that, if predictable, timing in the introduction of a new solution in an (electronic) market is fundamental: it should be postponed as late as possible to exploit current solutions but as soon as possible to make the critical mass inflection point occurring before the finite time singularity takes place.

A common point among the approaches is that the process of institutionalization contains *feedback loops*. Since homogeneous actors are more likely to interact, homogeneity can foster feedback loops to both diffuse change and balance it. Thus, homogeneity should be researched to allow institutionalization when a new solution has already been identified. For example, although the disruptive potential of quantum computing has been identified, with its "emerging new organizing logic and structure" (p. 2532), lack of actors' knowledge, budget for infrastructures, and shortage of experts can inhibit its institutionalization (Rietsche et al. 2022).

Finally, in dealing with the de/re-institutionalization needed for phase transitions, the dismantling of current institutional arrangements (de-institutionalization) is needed for a new set of coordinating activities across multiple actors, building new relationships with actors, and navigating cultural, normative, and even legal barriers. Re-institutionalization can take various forms, such as creating new regulatory frameworks, emergence of new professional bodies, establishing new forms of governance, or reorganizing new technological solutions. It creates opportunities for actors to engage in new and innovative ways and to develop more personalized and flexible institutions that meet the needs of the ecosystem's actors. This is the case of the current transformation of healthcare systems, boosted by the Covid-19 pandemic and digital technology advancements, that have de-institutionalized many practices and institutionalized others. However, the process has not brought the healthcare service ecosystem to new stability, and studies are trying to support researchers in capturing the current status of institutionalization of solutions and trying to forecast the future (Ostern et al. 2021).

Institutional work near the edge of chaos According to the American School concept of balance near the edge of chaos, a strategy for driving the service ecosystem towards a phase transition could consider the need to instill a change in institutional arrangements to promote a lack of satisfaction and understanding for agents and thus motivate them to enact change (Dooley 1997), while strengthening other institutional arrangements in place to loose not too much stability that could bring to collapse during the change. This type of institutional work (Koskela-Huotari et al. 2016) refers to "the purposive action of individuals and organizations aimed at creating, maintaining and disrupting institutions" (Lawrence and Suddaby 2006, p. 215). It is challenging in phase transitions according to which shifts are radical. Indeed, these disruptive activities most often occur within existing institutional arrangements (as actors cannot 'step outside' when trying to make a change). Furthermore, institutional work needs legitimacy in the service ecosystem for disruption (Thomas and Ritala 2022) to occur. As an ecosystem contains numerous actors guided by shared institutional arrangements, the legitimacy of the new and disrupted institutional arrangements is gained through a collective acceptance and institutionalization of these new institutions. This is often not straightforward and may imply contestations and conflicts (Gulati et al. 2012). For example, to develop standards for technologies and work practices in digital platform ecosystems, Costabile et al. (2022) suggested to: (i) developing new standards with the diffusion process in mind, so that the new solution itself can be considered beneficial for most of the target actors and likely adopted by them; (ii) engaging key actors in the process to enhance the possibilities of adoption and diffusion; (iii) leveraging extant standards and knowledge to keep at least in part the previous stability and build on it, trying to balance eventual drifts at the edge of chaos.

Thinking about the concept of self-organized criticality, for example, institutional work for driving a transition could be directed towards not waiting for the accurate development of a market strategy or the full understanding of other market actors' strategies before starting investments in a new technology. Indeed, the aggregation around a new technology could be already enough to acquire the critical mass to make the sandpile "flowing". A new avalanche of actors, then, will easily follow aggregating around the technology with a small uncertainty, and this amplifying feedback will bring new practices that, in turn, will provide the data to develop strategies and tactics around the technology.

Service ecosystem viability It also needs to be stated that phase transitions are central to the viability of a service ecosystem as the lack of change, according to complexity science, leads to stagnation and the system's death over time. However, a phase transition that lacks anchoring in new regulating institutional arrangements will drive the system to the edge of chaos. This is because a phase transition creates dynamic stability, forcing the actors to gain a better fit by avoiding the edge of chaos (as the three approaches suggest, respectively). Reconciled from an S-D logic perspective, indeed, the three approaches show that the viability of a service ecosystem is not related to a system in total harmony. Instead, viability can be related to crossing the edge of the complete order (first approach), moving into an area of complexity where order and chaos are balanced (second approach) by targeting institutional complexity, and trying to stay near the edge of chaos to renovate continuously, taking into account that institutional instability could also result in systems dissolution (third approach). Another way to keep the system alive is to actively drive change by creating disharmony in the system, such as instigating actors' dissatisfaction. These traditionally deviant behaviors can be central for the survivor of the system according to the three complexity science approaches. It also points toward the importance for actors to understand and act upon the feedback loops in the system. A relevant question from this insight is how little chaos is too little and how much chaos is too much for the system to handle.

References

Akaka, M.A., Koskela-Huotari, K., Vargo, S.L.: Formalizing service-dominant logic as a general theory of markets: taking stock and moving forward. AMS Rev. 1–15 (2021)

Allen, F., Gu, X., Jagtiani, J.: Fintech, cryptocurrencies, and CBDC: financial structural transformation in China. J. Int. Money Financ. **124**, 102625 (2022)

Alt, R.: From competitive advantage to social welfare. Electron. Mark. **32**, 487–492 (2022)

Anderson, P.: Complexity theory and organization science. Organ. Sci. **10**(3), 216–232. http://www.jstor.org/stable/2640328 (1999)

Argyris, C.: Overcoming organizational defenses: facilitating organizational learning. Allyn and Bacon (1990)

Arthur, W.B.: Increasing returns and path dependence in the economy. University of michigan Press (1994)

Arthur, W.B.: Foundations of complexity economics. Nat. Rev. Phys. **3**, 136–145 (2021)

Bak, P., Kan Chen.: Self-organized criticality. Sci. Am. 46–53 (1991)

Banh, L., Strobel, G.: Generative artificial intelligence. Electron Mark. **33**(63) (2023)

Blaschke, M., Riss, U., Haki, K., Aier, S.: Design principles for digital value co-creation networks: a service-dominant logic perspective. Electron. Mark. **29**, 443–472 (2019)

Brodie, R.J., Löbler, H., Fehrer, J.A.: Evolution of service-dominant logic: towards a paradigm and metatheory of the market and value cocreation? Ind. Mark. Manage. **79**, 3–12 (2019)

Cantwell, J., Dunning, J., Lundan, S.: An evolutionary approach to understanding international business activity: the co-evolution of MNEs and the institutional environment. J. Int. Bus. Stud. **41**, 567–586 (2010)

Capra, F.: The web of life: a new synthesis of mind and matter, p. 336. Flamingo, London (1996)

Cilliers, P.: Complexity and postmodernism. Understanding complex systems. Routledge, London (1998)

Colding, J., Barthel, S.: Exploring the social-ecological systems discourse 20 years later. Ecol. Soc. **24**(1), 2 (2019)

Costabile, C., Iden, J., Bygstad, B.: Building digital platform ecosystems through standardization: an institutional work approach. Electron. Mark. **32**, 1877–1889 (2022)
Pels, J.: Change of era or era of changes. In: Marinov, M.A., Marinova, S.T. (eds.) Covid-19 and international business: Change of era, pp. 19–25. Routledge (2020)
Dooley, K.J.: A complex adaptive systems model of organization change. Nonlinear Dyn. Psychol. Life Sci. **1**, 69–97 (1997)
Eidelson, R.J.: Complex adaptive systems in the behavioral and social sciences. Rev. Gen. Psychol. **1**(1), 42–71 (1997). https://doi.org/10.1037/1089-2680.1.1.42
Eldredge, N., Gould, S.J.: Punctuated equilibria: An alternative to phyletic gradualism. In: Schopf, T.J.M. (ed.) Models in paleobiology, pp. 82–115. Cooper&Co, San Francisco (1972)
Fuchs, C.: Dialectical philosophy and self-organization. In: Arshinov, V., Fuchs, C. (eds.) Causality, emergence, self-organization, pp. 195–244. NIA-Priroda, Moscow (2003)
Fui-Hoon Nah, F., Zheng, R., Cai, J., Siau, K., Chen, L.: Generative AI and ChatGPT: applications, challenges, and AI-human collaboration. J. Inf. Technol. Case Appl. Res. **25**(3), 277–304 (2023)
Fujita, S., Vaughan, C., Vargo, S.L.: Service ecosystem emergence from primitive actors in service dominant logic: an exploratory simulation study. In: Proceedings of the 51st Hawaii International Conference on System Sciences (2018)
Fujita, S., Vaughan, C., Vargo, S.L: Service ecosystems emergence and interaction: a simulation study. In: Proceedings of the 52nd Hawaii International Conference on System Sciences (2019)
Gell-Mann, M.: The Quark and the Jaguar. Freeman, New York (1994)
Giddens, A.: The Constitution of Society. Polity Press, Cambridge (1984)
Giesler, M., Fischer, E.: Market System Dynamics. Marketing Theory **17**(1), 3–8 (2017)
Glansdorff, P., Prigogine, I.: Thermodynamic Theory of Structure, Stability, and Fluctuations. Wiley-Interscience, New York) (1971)
Greenwood, R., Hinings, C.R.: Understanding radical organizational change: bringing together the old and the new institutionalism. Acad. Manag. Rev. **21**(4), 1022–1054 (1996)
Gulati, R., Wohlgezogen, F., Zhelyazkov, P.: The two facets of collaboration: cooperation and coordination in strategic alliances. Acad. Manag. Ann. **6**(1), 531–583 (2012)
Gunderson, L.H., Holling, C.S Panarchy: Understanding transformations in human and natural systems. Island press (2002)
Haken, H.: Synergetics, an introduction: non-equilibrium phase transitions and self-organizationin physics, chemistry, and biology, 3rd edn. Springer-Verlag, Berlin (1983)
Hatum, A., Pettigrew, A.M.: Determinants of organizational flexibility: a study in an emerging economy. Br. J. Manag. **17**(2), 115–137 (2006)
Holbrook, M.B (2003) Adventures in complexity. Acad. Mark. Sci. Rev. **06**
Holland, J.: Complexity: A very short introduction. Oxford University Press, Oxford (2014)
Howe, M.L., Lewis, M.D.: The importance of dynamic systems approaches for understanding development. Dev. Rev. **25**(3–4), 247–251 (2005)
Jantsch, E.: The self-organizing universe: scientific and human implications of the emerging paradigm of evolution. Pergamon (1980)
Kaartemo, V., Nyström, A.G.: Emerging technology as a platform for market shaping and innovation. J. Bus. Res. **124**, 458–468 (2021)
Kaartemo, V., Kowalkowski, C., Edvardsson, B.: Enhancing the understanding of processes and outcomes of innovation: the contribution of effectuation to SD logic. In: Sage handbook of service-dominant logic, pp. 522–535 (2018)
Kauffman, S.A.: (1969). Metabolic stability and epigenesis in randomly constructed genetic nets. J. Theor. Biol. **22**, 437–467 (1969)
Kauffman, S.A.: The origins of order: self organization and selection in evolution. Oxford University Press, New York (1993)
Kauffman, S.A.: At home in the universe: the search for laws of self-organization and complexity. Oxford University Press, New York (1995)

Kijima, K., Toivonen, M., Ruutu, S.: Service Ecosystems Innovation in Systemic Perspective: Transitions and Coevolutions. In: Toivonen M. (ed.) Service Innovation, Translational Systems Sciences vol. 6, pp. 51–67 (2016)

Kim, W.C., Mauborgne, R.: Value innovation. Harv. Bus. Rev. **82**(7/8), 172–180 (2004)

Kleinaltenkamp, M., Corsaro, D., Sebastiani, R.: The role of proto-institutions within the change of service ecosystems. J. Serv. Theory Pract. **28**(5), 609–635 (2018). https://doi.org/10.1108/JSTP-12-2017-0241

Koskela-Huotari, K., Edvardsson, B., Jonas, J.M., Sörhammar, D., Witell, L.: Innovation in service ecosystems—breaking, making, and maintaining institutionalized rules of resource integration. J. Bus. Res. **69**(8), 2964–2971 (2016)

Kraus, S., Jones, P., Kailer, N., Weinmann, A., Chaparro-Banegas, N., Roig-Tierno, N.: Digital transformation: an overview of the current state of the art of research. SAGE Open **11**(3) (2021)

Lawrence, T. B., Suddaby, R.: 1.6 institutions and institutional work. In: The Sage handbook of organization studies, pp. 215–254 (2006)

Li, F.: Leading digital transformation: three emerging approaches for managing the transition. Int. J. Oper. Prod. Manag. **40**(6), 809–817 (2020)

Lichtenstein, B.B., Plowman, D.A.: The leadership of emergence: a complex systems leadership theory of emergence at successive organizational levels. Leadersh. q. **20**(4), 617–630 (2009)

Lieberman, M.B., Montgomery, D.B.: First-Mover Advantages. Strat. Manag. J. **9**(S1), 41–58 (1988)

Lintula, J. M., Päivärinta, T., Tuunanen, T.: Value Co-creation for Smart Villages: The Institutionalization of Regional Service Ecosystems. In: International Conference on Information Systems. Association for Information Systems (2020)

Lusch, R.F., Vargo, S.L.: Service-Dominant Logic: Premises Possibilities. Cambridge University Press, New York, Perspectives (2014)

MacInnis, D.J.: A framework for conceptual contributions in marketing. J. Mark. **75**(4), 136–154 (2011)

MacIntosh, R., MacLean, D.: Conditioned emergence: a dissipative structures approach to transformation. Strateg. Manag. J. **20**(4), 297–316 (1999)

Maguire, S., McKelvey, B., Mirabeau, L., Öztas, N.: 1.5 complexity science and organization studies. In: Sage handbook of organization studies, pp. 165–214 (2006)

Mainzer, K.: Thinking in complexity: the complex dynamics of matter, mind, and mankind, 3rd edn. Springer-Verlag, New York (1997)

March, J.G., Olsen, J.P.: Rediscovering institutions: the organizational basis of politics. Free Press, New York (1989)

Markose, S.M.: Computability and evolutionary complexity: markets as complex adaptive systems (CAS). Econ. J. **115**(June), F159–F192 (2005)

Mathews, K.M., White, M.C., Long, R.G.: Why study the complexity sciences in the social sciences? Hum. Relat. **52**(4), 439–462 (1999)

McKelvey, B.: What is complexity science? It is really order-creation science. Emerg. J. Complex. Issues Organ. Manag. **3**(1), 137–157 (2001)

Meynhardt, T., Chandler, J.D., Strathoff, P.: Systemic principles of value co-creation: synergetics of value and service ecosystems. J. Bus. Res. **69**(8), 2981–2989 (2016)

Nenonen, S., Gummerus, J., Sklyar, A.: Game-changers: dynamic capabilities' influence on service ecosystems. J. Serv. Manag. **29**(4), 569–592 (2018)

Nenonen, S., Kjellberg, H., Pels, J., Cheung, L., Lindeman, S., Mele, C., Sajtos, L., Storbacka, K.: A new perspective on market dynamics: market plasticity and the stability–fluidity dialectics. Mark. Theory **14**(3), 269–289 (2014)

Oliver, C.: The antecedents of deinstitutionalization. Organ. Stud. **13**(4), 563-588 (1992)

Oschinsky, F.M., Klein, H.C., Niehaves, B.: Invite everyone to the table, but not every course. Electron. Mark. **32**, 1925–1941 (2022)

Ostern, N., Perscheid, G., Reelitz, C., et al.: Keeping pace with the healthcare transformation: a literature review and research agenda for a new decade of health information systems research. Electron. Mark. **31**, 901–921 (2021)

Pascale, R., Milleman, M., Gioja, L.: Surfing the edge of chaos: the laws of nature and the new laws of business. Currency (2001)

Pels, J., Mele, C.: Is service management experiencing a change of era? In: The Palgrave Handbook of Service Management, pp. 71–84. Springer International Publishing, Cham (2022)

Peñaloza, L., Mish, J.: The nature and processes of market co-creation in triple bottom line firms: Leveraging insights from consumer culture theory and service dominant logic. Mark. Theory **11**(1), 9–34 (2011)

Peters, B.G.: Institutional theory in political science: the "new institutionalism. Pinter, London (1999)

Peters, L.D.: Heteropathic versus homopathic resource integration and value co-creation in service ecosystems. J. Bus. Res. **69**(8), 2999–3007 (2016)

Peters, L.D., Nenonen, S., Polese, F., Frow, P., Payne, A.: Viability mechanisms in market systems: prerequisites for market shaping. Journal of Business & Industrial Marketing **35**(9), 1403–1412 (2020)

Phillips, M.A., Ritala, P.: A complex adaptive systems agenda for ecosystem research methodology. Technol. Forecast. Soc. Chang. **148**, 119739 (2019)

Pohlmann, A., Kaartemo, V.: Research trajectories of Service-Dominant Logic: Emergent themes of a unifying paradigm in business and management. Ind. Mark. Manage. **63**, 53–68 (2017)

Polese, F., Payne, A., Frow, P., Sarno, D., Nenonen, S.: Emergence and phase transitions in service ecosystems. J. Bus. Res. **127**, 25–34 (2021). https://doi.org/10.1016/j.jbusres.2020.11.067

Polese, F., Sarno, D., & Vargo, S. L. (2020). *The role of emergence in service systems*. In Proceedings of the 53rd Hawaii International Conference on System Sciences.

Preiser, R., Biggs, R., De Vos, A., & Folke, C. (2018). Social-ecological systems as complex adaptive systems. *Ecology and Society*, *23*(4).

Prigogine, I., Stengers, I.: Order out of chaos. Bantam Books, New York (1984)

Rietsche, R., Dremel, C., Bosch, S., et al.: Quantum Computing. Electron Mark. **32**, 2525–2536 (2022). https://doi.org/10.1007/s12525-022-00570-y

Rogers, E.M.: Diffusion of innovation. Free Press, New York, NY (2003)

Rogers, E., Medina, U.E., Rivera, M.A., Wiley, J.C.: Complex adaptive systems and the diffusion of innovations. Public Sect. Innov. J. **10**(3) (2005)

Sarasvathy, S.D.: Causation and effectuation: toward a theoretical shift from economic inevitability to entrepreneurial contingency. Acad. Manag. Rev. **26**(2), 243–263 (2001)

Sarno, D., Siano, P.: Exploring the adoption of service-dominant logic as an integrative framework for assessing energy transitions. Sustainability **14**(15), 1–26 (2022). https://doi.org/10.3390/su14159755

Schöbel, S.M., Leimeister, J.M.: Metaverse platform ecosystems. Electron. Mark. **33**, 12 (2023)

Senge, P.: The fifth discipline: the art and practice of the learning organization. Century Business, London (1990)

Siltaloppi, J., Koskela-Huotari, K., Vargo, S.L.: Institutional complexity as a driver for innovation in service ecosystems. Serv. Sci. **8**(3), 333–343 (2016)

Smith, C., Gemmill, G.: Change in the small group: a dissipative structure perspective. Hum. Relat. **44**(7), 697–716 (1991). https://doi.org/10.1177/001872679104400704

Taillard, M., Peters, L.D., Pels, J., Mele, C.: The role of shared intentions in the emergence of service ecosystems. J. Bus. Res. **69**(8), 2972–2980 (2016)

Taleb, N.N.: Antifragile: things that gain from disorder. Penguin (2012)

Teece, D.J.: Explicating dynamic capabilities: the nature and microfoundations of (sustainable) enterprise performance. Strateg. Manag. J. **28**(13), 1319–1350 (2007)

Teece, D.J.: Dynamic capabilities and entrepreneurial management in large organizations: toward a theory of the (entrepreneurial) firm. Eur. Econ. Rev. **86**, 202–216 (2016)

Thomas, L.D., Ritala, P.: Ecosystem legitimacy emergence: a collective action view. J. Manag. **48**(3), 515–541 (2022)
Tuominen, T., Edvardsson, B., Reynoso, J.: Institutional change and routine dynamics in service ecosystems. J. Serv. Mark. **34**(4), 575–586 (2020)
Vargo, S.L., Lusch, R.F.: Evolving to a new dominant logic for marketing. J. Mark. **68**, 61–17 (2004)
Vargo, S.L., Lusch, R.F.: Service-dominant logic; further evolution. J. Acad. Mark. Sci. **36**(1), 1–10 (2008)
Vargo, S.L., Lusch, R.F.: Service-dominant logic 2025. Int. J. Res. Mark. **34**(1), 46–67 (2017)
Vargo, S.L., Peters, L., Kjellberg, H., Koskela-Huotari, K., Nenonen, S., Polese, F., Sarno, D., Vaughan, C.: Emergence in marketing: an institutional and ecosystem framework. J. Acad. Mark. Sci. **51**(1), 2–22 (2023a)
Vargo, S.L., Wieland, H., O'Brien, M.: Service-dominant logic as a unifying theoretical framework for the re-institutionalization of the marketing discipline. J. Bus. Res. **164** 113965 (2023)
Vargo, S.L.: Conceptual reconciliation for clarity and impact. AMS Rev. **13**, 169–172 (2023)
Vargo, S.L., Lusch, R.F.: Institutions and axioms: an extension and update of service-dominant logic. J. Acad. Mark. Sci. **44**(1), 5–23 (2016)
Vargo, S.L., Akaka, M.A., Wieland, H.: Rethinking the process of diffusion in innovation: a service-ecosystems and institutional perspective. J. Bus. Res. **116**, 526–534 (2020)
Vial, G.: Understanding digital transformation: a review and a research agenda. J. Strateg. Inf. Syst. **28**(2), 118–144 (2019)
Von Bertalanffy, L.: The history and status of general systems theory. Acad. Manag. J. **15**(4), 407–426 (1972)
Waldrop, M.M.: Complexity: the emerging science at the edge of order and chaos. Open Road Integrated media (1992)
Walker, B., Holling, C.S., Carpenter, S.R., Kinzig, A.: Resilience, adaptability and transformability in social–ecological systems. Ecol. Soc. **9**(2) (2004)
Weick, K.E.: Introductory essay—Improvisation as a mindset for organizational analysis. Organ. Sci. **9**(5), 543–555 (1998)
West, G.B.: Scale: the universal laws of growth, innovation, sustainability, and the pace of life in organisms, cities, economies, and companies. Penguin Press, New York (2017)
Wiltbank, R., Dew, N., Read, S., Sarasvathy, S.D.: What to do next? The case for non-predictive strategy. Strateg. Manag. J. **27**(10), 981–998 (2006)

Chapter 2
Using a Systems Perspective for Sharing Market Views and Changes for Market Shaping

Debora Sarno, Francesco Polese, and Antonietta Megaro

Abstract Collaborative market shapers should be aligned to increase market viability. Leveraging systems theory and systems thinking, we suggest considering the market structure and system in order to derive specific market shaping strategies and collectively enact them. We contribute to market shaping literature by: (i) delineating the elements of the market view to be shared, and (ii) differentiating the types of market change, providing a typology based on the extent of change measured in terms of changes to market problems, solutions, and structures. This work can support researchers and collaborative shapers in analyzing and designing shared market views and strategies taking into account the implications in terms of the degree of predictability and manageability of the systemic outcomes of shaping.

Keywords Market system · Market transformation · Market reformation · Market emergence · Market design · Market system dynamics

Introduction

Multiple current streams of literature perceive markets as complex adaptive systems (Mitleton-Kelly 2003) that can produce novelty and surprises (Markose 2005; Giesler and Fischer 2017). In such view, markets are comprised of a variety of actors (not only customers but also companies, third-party organizations, authorities, etc.) that offer institutionalized solutions to other actors' needs (Vargo and Lush 2014) and enact multiple practices (Kjellberg et al. 2012) for co-creating value. In this view, every

D. Sarno (✉)
University of Naples Parthenope, Naples, Italy
e-mail: debora.sarno@uniparthenope.it

F. Polese · A. Megaro
University of Salerno, Fisciano, Italy
e-mail: fpolese@unisa.it

A. Megaro
e-mail: amegaro@unisa.it

D. Sarno et al. (eds.), *Complexity and Emergence in Market Ecosystems*, New Economic Windows, https://doi.org/10.1007/978-3-031-81942-1_2

market actor feels an increasing need to make sense of market networks, practices, symbols, rules, and beliefs, which are prerequisites to make a strategical choice related to driving or being driven by markets (Jaworski et al. 2000; Kumar et al. 2000). The existing concept of markets has been recently revised and deepened by the research stream on the empirical phenomenon of market shaping, according to which market shapers (individual actors or collectives) can exploit market plasticity to systemically alter the market (Nenonen et al. 2020). The purpose underlying market shaping can be, for example, fostering social change (Kullak et al. 2022), managing threats (e.g., new market regulations, aggressive actions of other shapers, lack of resources), or creating new opportunities (Nilsson et al. 2023), especially when old market models are already being challenged, as seen in the vaccine market during the COVID-19 pandemic (Nenonen and Storbacka 2020).

Market multiplicity—the existence of multiple market views on a market due to actors' different market perspectives, actors' actions based on different definitions of the relevant environment, and actors' market-making activities—can impact the market (Kjellberg and Helgesson 2006) and serve as an important source of novelty and creativity for shapers. However, since shapers need to collaborate with other actors (Nenonen and Storbacka 2020) to reach an agreement on and pursue a shaping strategy, conflicting market views (Kjellberg et al. 2012) can become risky and potentially destructive over time. For example, a shaper who is unaware of the current market relationships of other collaborators, may accidentally act against his own "team and game" by disclosing key information to competitors. This is worsened by the fact that "collaborators" may have their own objectives and new, unpredictable properties can emerge (Nenonen et al. 2019; Vargo et al. 2023). Moreover, highly improbable events (Taleb 2008) may occur that potentially generate new markets (Struben et al. 2020; Beninger and Francis 2021).

A systemic understanding of markets and their change dynamics is crucial to collaborate for successful market shaping (Flaig et al. 2021), and reduce market multiplicity in times of coordinated actions. We contribute to this understanding by supporting managers in sharing viewpoints while being aware of the uncertainties related to the extent of change they attempt to pursue. In particular, we delineate what should be shared to increase market viability (Peters et al. 2020), such as the actual market (the "AS IS market view") and the envisioned shaped market ("TO BE market view") views (which could be practically agreed on based on Delphi methods, Nyström and Kaartemo 2022). Moreover, we suggest that, by comparing AS IS and TO BE market views defining a specific shaping strategy, shapers should be able to recognize the extent of change (or change type) they must instill in the market, the elements that must be changed, and the possible outcomes (Fig. 2.1). For such reason, we differentiate among market systems changes providing a typology.

The remainder of our paper is structured as follows. Section 2.2 describes the research design used to develop the conceptualization. Section 2.3 presents the theoretical background for our study. In Sect. 2.4 and 2.5 our delineating and differentiating contributions are provided. Section 2.6 closes with theoretical and practical implications.

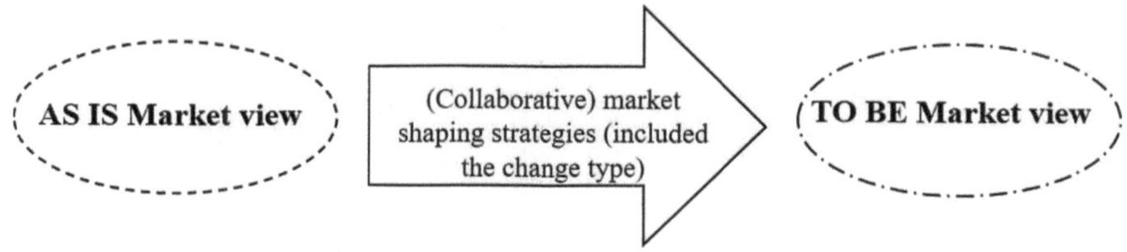

Fig. 2.1 Relationship between market view(s) and market shaping strategy to be shared by collaborators for market shaping. *Source* Authors' elaboration

Research Design

In this section, we show how we meet our research goals. First, we built the theoretical background by reviewing the market shaping (domain theory, MacInnis 2011) literature to ascertain the meaning of a market view and the reasons for sharing it. In particular, we performed a systematic literature review (Nightingale 2009; Mengist et al. 2020) on the topic using Scopus. For the first papers' extrapolation, we used the keywords "market AND shaping AND view" and "market AND shaping AND change." To filter out non-useful results, we excluded duplicate papers, papers referring to other topics, and papers for which the full text was unavailable. We obtained 88 articles, which were subsequently analyzed through a conceptual analysis of the data with the aim of extracting the main concepts and highlighting their relevance concerning our topic. Sections 3.1 and 3.2 present the results.

Next, given that markets can be considered (complex adaptive) systems, we propose systems theories and systems thinking as a theoretical lens (method theory, Lukka and Vinnari 2014) for a novel way of seeing, understanding, and observing market views and market systems changes to pursue effective market shaping strategies. Thus, we leverage MacInnis's (2011) seminal work on conceptual papers in marketing to develop a conceptual contribution to the market shaping literature with the goals of delineating and differentiating.

The *delineating* conceptual goal allows researchers to detail, articulate, trace, describe, and/or represent a given phenomenon by mapping the relationships and components of this construct and developing a conceptual framework to better understand it. We specifically delineate the market view to be shared among shapers to provide a simple way of practically defining it (i.e., the "market structure and system").

The *differentiating* conceptual goal allows researchers to provide conceptual progress and new insights into a specific phenomenon by distinguishing, analyzing, sizing, and/or classifying it through typologies, taxonomies, and classifications. We specifically differentiate the market systems change to support researchers and practitioners in both envisioning and analyzing actual changes.

Figure 2.2 shows our approach, adding details on the method theory and contribution described later in the chapter.

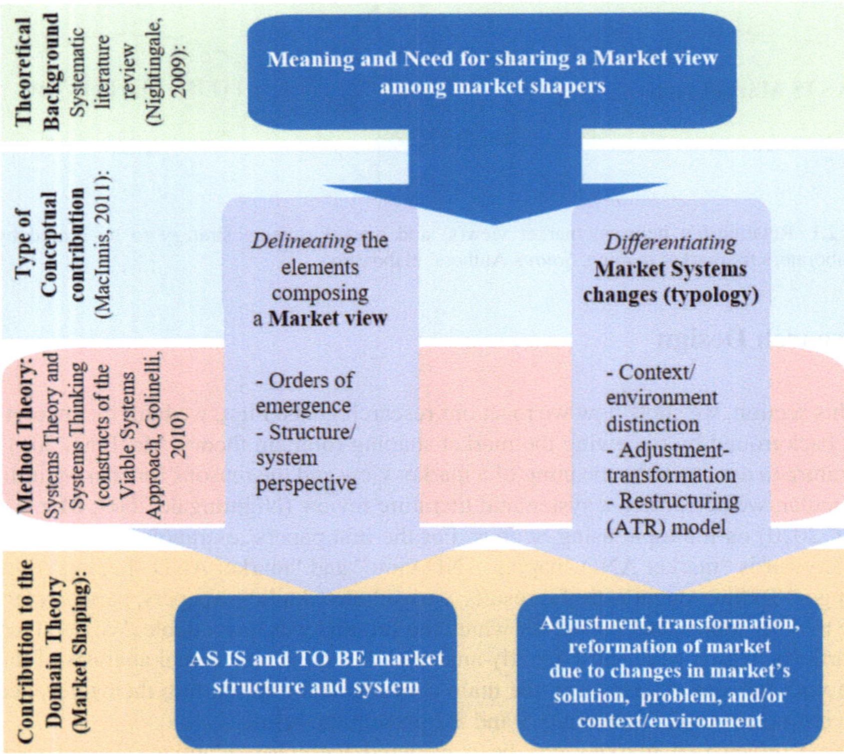

Fig. 2.2 Synthesis of the methodology and contributions of the manuscript. *Source* Authors' elaboration

Theoretical Background

The Need for Sharing a Market View

Market shapers should measure their resources because initiatives without enough followers have insufficient support to be successful (Nenonen and Storbacka 2020). Specifically, when leveraging the idea of a minimum viable product (from product development) to shape the market, scholars have proposed involving a subset of market actors, called the minimum viable system (MVS), which is "the smallest and simplest system that can support any market shaping vision … [taking into account] division of labor and distribution of relevant knowledge … dependence on common resources … and common rules of the game …" (Nenonen and Storbacka 2020, p.269). Indeed, market shaping is a collective endeavor (Baker and Nenonen 2020; Struben et al. 2020; Beninger and Francis 2021) that depends on multiple actors' joint engagement in institutional processes (Kleinaltenkamp et al. 2021).

Shared intentions, which are preconditions for action, have been also identified as responsible for service ecosystems (Taillard et al. 2016) as well as market formation (Hawa et al. 2020). A shared understanding of a particular market can enhance its stability (Fligstein 2001) and may enable collaborators to "learn as a group" (Nenonen and Storbacka 2020).

Moreover, market shapers must reduce market multiplicity (Kjellberg and Murto 2021). Although it is generally a source of novelty and creativity, market multiplicity must be reduced (Lichtenstein et al. 2006). Similarly, awareness of the values and institutional logic that influence collaborators (Kaartemo et al. 2020), the existence of joint problems, and similarities in size, values, and business objectives (Baker and Nenonen 2020) can be useful for shaping.

Finally, the literature highlights that since markets are complex adaptive systems (Nenonen et al. 2019), the outcome of market shaping is not always predictable (Layton and Duffy 2018; Vargo et al. 2023), making the sharing of market mapping and strategies much more critical (Möller et al. 2020), offering elements to try to make sense of novelty in the broader framework of the envisioned outcomes of market shaping (Fehrer et al. 2020).

The Meaning of a "Market View"

Different theories and perspectives have dealt with market system dynamics (Pedeliento et al. 2023) and market shaping (Nenonen and Storbacka 2021). Thus, diverse ideas on the definition of a "market view", that is, a "TO BE" view (Storbacka and Nenonen 2011) have been developed, and diverse strategies to shape markets have been proposed accordingly. Thus, market shapers should influence other market actors to share their market view.

According to an institutional perspective, markets are conceptualized as institutionalized solutions to actors' needs or problems (Vargo and Lusch 2014; Lusch and Vargo 2014; Akaka et al. 2021). According to this perspective, a market change represents the emergence and institutionalization of new solutions (Vargo et al. 2015), or, in other words, solutions that become intersubjective realities. Therefore, a market view that is envisioned and shared is a strategic template (Zietsma and McKnight 2009) leading to collective market shaping (Baker and Nenonen 2020). Complementing this, but with a practice-driven institutionalism perspective, a process model of engagement-driven institutionalization consists of three stages: (1) the synchronization of collective dispositions and behaviors, (2) the development of temporal stability of behaviors through self-reinforcing mechanisms resulting in practices, and (3) the ensuing institutional change for market shaping (Kleinaltenkamp et al. 2021).

With a broader and more strategic perspective, shapers could imagine an alternate state of the market (a protovision) and communicate it (Patvardhan and Ramachandran 2020; Nenonen and Storbacka 2020). In doing so, shapers would author new meanings (Weick 1995) and be able to articulate simplified versions of the protovision "through some medium that enables boundary crossing between market actors"

(Nenonen and Storbacka 2020, p. 268). Thus, similarly to what is described in the collective market shaping process (Baker and Nenonen 2020), and given that a vision should be scalable so that a critical mass could be achieved, its development will require moving from a protovision to a declaration of intent that can guide strategy development. Finally, as aforementioned, several authors have recognized that a shared understanding of the market can enhance its stability.

Informing the Market View with a Systems View

Market shaping strategies are implemented in markets intended as complex adaptive systems (Nenonen et al. 2019; Diaz Ruiz et al. 2020) that are characterized by the principles of self-organization and emergence (Markose 2005). Accordingly, shapers' design capabilities can be expressed at a micro level of market analysis (Windhal et al. 2020), but there is no external organizing force. Furthermore, considering markets as systems allows us to take into account that they are characterized by a non-linear relationship between actors' actions and market changes (Mele et al. 2021), which constitutes "an important missing link in both the scholarly and managerial dialogue on market shaping strategies" (Peters et al. 2020, p.7). Indeed, emerging market properties cannot be understood by studying their parts in isolation but only by shifting attention from the individual parts to the whole (Checkland 1997).

The systems turn in marketing has quite a long story, moving from the initial thoughts of Alderson (1965) and Lewis and Erickson (1969), who developed a general theory of marketing, followed by applications to micro and macro marketing systems (Fisk 1967) and then the integration of marketing systems in society (Dixon 1967). The systems research on markets and marketing phenomena has successfully introduced complex adaptive systems to marketing (Holbrook 2003), supplemented with further studies by Layton (2007; 2014). Based on a systems view (Capra 1996), markets are observed by continuously moving back and forth from the whole to the parts and vice versa (Vargo et al. 2017, 2023). In particular, a systems view has been identified as a way for companies to look beyond the seller–buyer dyads and search for relationships with actors willing to co-create value in a more extensive system, particularly for market shaping (Möller et al. 2020; Peters et al. 2020; Kullak et al. 2022). This approach offers the possibility of understanding the evolutionary processes driving change in ideas, behaviors, and social practices (Layton 2014) and overcoming the limitations of a solid reductionist view in which the focus is on single actors and transactions as the only link between companies and environments (Tronvoll et al. 2018). Above all, a systems view can enable different forms of marketing reasoning by highlighting the dynamics of market systems in managing an increasingly complex institutional environment. Examples include exploiting new technologies (Layton 2011) and suggesting behavioral changes that favor iterative and co-creative change processes in order to resolve complex problems (Domegan et al. 2016). Furthermore, as Peters et al. (2020) pointed out, a systems view can facilitate reflections on market shaping related to destabilizing, changing, and restabilizing markets. Such a view is needed not only by academics but also by practitioners, who mainly rely on static market views (Wieland et al. 2021). In other words, to increase the theoretical work's relevance, academics and practitioners must be encouraged to develop a more holistic and reflexive understanding of markets (Wieland et al. 2021; Möller et al. 2020; Vargo et al. 2017; Mele et al. 2015).

We acknowledge that service-dominant logic is a systems-based general theory of markets (Akaka et al. 2021) that should be embraced especially to leverage its multi-level perspective (Chandler and Vargo 2011) in analyzing markets, from the dyadic interactions of actors of the micro-level to the macro/mega-level perspectives of other systems and societies with which markets are intertwined. However, we notice that a deep understanding of market system changes has yet to be developed in either service-dominant logic or in other market theories. Thus, we propose adopting an overarching perspective of "market systems" to allow shaping actors to share market views and the extent of change. Applications of other market theories may be needed to develop analyses and detailed strategies for market shaping.

To build the market view to be shared with the MVS, we integrate the market systems view (Möller et al. 2020) that originates from multiple studies, primarily Mele et al. (2015), with insights from both systems theory and systems thinking. While the former illustrates how structures come into being, the latter addresses how different structures have different dynamic consequences giving rise to varying behaviors (Senge 2001). In particular, the viable systems approach (VSA) (Golinelli 2010; Barile and Polese 2010)—a meta-theory rooted in systems thinking that provides an essential body of knowledge for the management of socioeconomic systems—has been developed to offer a comprehensive systems view and an integrated approach to management (Mele et al. 2010; Barile and Polese 2010; Badinelli et al. 2012;) and marketing (Golinelli et al. 2012; Barile et al. 2012). It introduces the concepts of actors' perspectives and engagement in complex (adaptive) systems (Badinelli et al. 2012), actors' structural consonance and systemic resonance in systems (Wieland et al. 2012), and systems adaptation and viability (Polese et al. 2017). Moreover, it has contributed (Brodie et al. 2018) to service-dominant logic (Barile and Polese 2010; Wieland et al. 2012; Pels et al. 2014; Barile et al. 2016), which, in turn, has provided the conceptualization of the service ecosystem (Vargo and Lusch 2019) to make sense of market, societal, and organizational exchanges.

Delineating a "Unifying" Market View: Markets as Structure and Systems

We contribute to the market shaping domain by *delineating* the concept of a "market view." As suggested by MacInnis (2011), we adopt deductive reasoning, which enables us to benefit from a theories-in-use approach to describe and represent a specific entity, framing it within the broader conceptual world that surrounds it, and thus permitting a better understanding of the term.

Emergence and the AS IS/TO BE Market Views

The systems view, which allows for effectively tackling complexity in markets, considers the phenomenon of emergence (Vargo et al. 2023). We draw on Vargo et al.'s (2023) four orders of emergence, noting that the fourth order relates to marketing shaping since it focuses on the purposive work to create, maintain, and change institutions. In such an order, actors are aware of market properties and their roles in producing them by envisioning new potential service ecosystem properties and reflexively engaging in resource integration and service-for-service exchange to obtain newly designed emergent properties. Thus, taking into account such "awareness" of market properties, we posit that it can be achieved—and possibly shared—by "mapping" the market before envisioning its changes. Thus, we suggest identifying—and sharing—an "AS IS market view" when starting to collaborate as an MVS and collectively developing and achieving a "TO BE market view" through market shaping (Fig. 2.3). As shown in Fig. 2.1, the shaping strategy will link the AS IS and the TO BE market view. However, it should be noted that market shapers could be aware of the uncertainties of market dynamics due to emergence, as we discuss later in Sect. 2.5.

A Structure-System Perspective on Markets

Market shaping is a deliberate market strategy (Elg et al. 2012) according to which market shapers can leverage market plasticity, which is "a market's ability to take and retain form… '[F]orm' in a market context is not limited to structure alone but involves both structural and 'functional' aspects." (Nenonen et al. 2014, p. 5). Thus, both structural (characteristics) and systemic (dynamic, behavioral) views are highlighted as the focus of market shaping (Peters et al. 2020). The market structure comprises the set of market actors and their roles in the market (Jarowski et al. 2000). Goslin et al. (2017) describe the market boundaries with different categories of practices. Thus, markets could be modified by changing actors (a construction/deconstruction approach) or their roles (a functional-modification approach). In those cases, the focus of research can be on "market-as-noun" (Mele et al. 2015), wherein the investigated themes are the market entities (actors and relationships, i.e., the market structure) and market representations (the ways in which markets are signified).

In contrast, market behavior refers to the behaviors of all market actors, which can be modified directly (building/removing customers/competitor constraints) or indirectly by changing actors' mindsets (creating new/reversed customer/competitor preferences) (Jarowski et al. 2000). With a market-as-verb focus (Mele et al. 2015), the themes of interest are market performance (the actions taking place in the market, i.e., the market system) and market sense-making (market dynamics in terms of emergence and evolution).

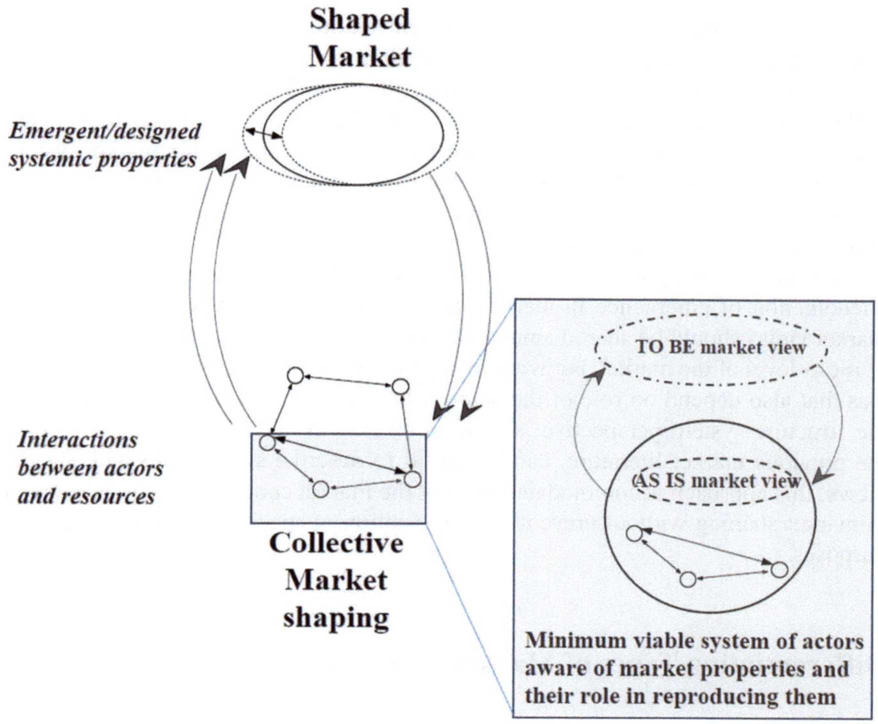

Fig. 2.3 Collective market shaping. The upward and downward arrows represent the interactions between actors, resources, and emergent properties. *Source* Authors' elaboration based on the fourth order of emergence of Vargo et al. (2023)

Similarly to these classifications, VSA suggests that each socio-material system can be analyzed through structural and dynamic perspectives (i.e., the structure-system paradigm) (Barile and Saviano 2011), which are two interrelated ways of observing systems (Barile and Polese 2010). The structural perspective provides the "means and mechanisms" by which market system actors operate (Checkland 1999, p. 102) following a reductionist viewpoint. Such a structural perspective incorporates micro, meso, and macro levels of market analysis to understand how they are interlinked (Möller et al. 2020). However, the structural perspective on a market system is insufficient to explain "what it does" (Beer 1979, 1985).

A systems perspective on markets considers the interaction of components according to the activation of structural relationships (Barile and Polese 2010), thus taking into account the market's dynamic aspect. Such systemic properties of the market differ from those of the structure's components (Checkland 1999). Thus, understanding the whole may not require understanding the parts (a holistic viewpoint). However, although some systems' properties are emergent and not reducible to components (Vargo et al. 2023; Kim 1999; Checkland 1999), others can be strictly related to the components' properties and their relationships. In other words, in the case of changes in market structure, whatever they may be, actors' self-organization may determine new systemic emergent properties (Vargo et al. 2023).

The structure-system perspective helps bridge the gap between reductionism and holism (Barile and Polese 2010). Notably, the structural aspects of a market system might be ex-ante designable. However, due to dynamics and emergence, the structure's evolutions and the market system's behavioral aspects can only be ex-post understandable (Golinelli 2010).

In sum, the adoption of a systems view and the acknowledgment of the phenomenon of emergence in markets allows us to posit that AS IS and TO BE market views should be shared among shapers, always setting as target the analysis of meso-level of the market, but without neglecting the other possible levels of analysis that also depend on role of the actor in the market. Moreover, the adoption of the structure-system perspective, also used by Pels et al. (2023) to accommodate the previous market literature, can be useful to describe such AS IS and TO BE views; this approach accommodates most of the market conceptualizations adopted in market shaping without preventing the adoption of specific theories for specific purposes.

Differentiating Types of Market Changes

We contribute to the market shaping domain by *differentiating* the diverse types of market changes that a market, intended as a complex (adaptive) systems, can manifest. In doing so, we develop a contribution coherent with the adulthood of the concept of "market system dynamics" (Pedeliento et al. 2023; Ozdamar Ertekin et al. 2020). As MacInnis (2011) suggested, we adopt analytical reasoning based on diverse literature sources to organize the relevant literature into types (Jaakkola 2020).

The issue of determining the extent of change of a market shaping initiative to be shared among shapers requires a comprehensive analysis. To contribute to the market shaping stream of research, we develop a typology of market changes for market shaping. In such a way, we suggest market shapers share not only the structural and systemic aspects to be changed in the AI IS market view to achieve the TO BE market view but also their awareness of the extent of change and its systemic and expected outcomes. Our approach offers the advantage of systematically thinking about desired changes in structures and their systemic consequences without forgetting possible unintended emergent properties due to complexity. As a metaphor, nowadays, more than ever, market shapers should share how to trigger artificial waves of change in their markets without forgetting that they could be required to surf through those and other possible waves due to unpredictable weather.

By enlarging our perspective, we do not limit the analysis of the market change to the market itself, but we also include the market environment. Indeed, markets are intrinsically dynamic as they are influenced—and influence—the institutional and knowledge environments in which they behave (Layton 2011). Leveraging VSA again, we differentiate the actual or specific market structure from the environment (thus deepening the structural aspect of the structure-system perspective) as follows.

- The environment can be seen as an extended structure that represents the set of all possible components (actors and resources) and the potential relationships among them.
- The actual or specific market structure embodies the configuration of components and the specific relationships between them selected among the alternatives of the environment.

In dealing with such a selection decision, every piece of information depends on the observer's perception and cannot be separated from them (von Glasersfeld 2001; Barile and Polese 2010). Thus, if the observers are markets' actors, the extent to which the selection of components and relationships is shared determines the common field of actors' understanding; that is, the sharing of interpretation schemes (Barile 2009b). This also implies a significant reduction in the multiplicity of potential partial market versions (Kjellberg et al. 2012), thereby increasing the likelihood that actors share common aims with regard to changes.

Based on a systems view of change, different extents—a typology—of market system change can be distinguished via escalation logic, adapting the adjustment-transformation-restructuring (ATR) model of the VSA (Barile 2009a; Golinelli 2010) related to organizational change. Dealing with the "translation" of systems purpose to market systems, we concur with Vargo and Lusch's (2014) statement that markets are institutionalized solutions to problems. Thus, we view adjustments and transformations as market changes in which the problems to solve are the same but resolved in slightly different ways. In particular, reformulation is related to (i) finding radically new solutions to problems or (ii) changing the problem to be addressed. Hauser et al. (2006) support the former, claiming that "[b]y finding new solutions to problems, innovation can destroy existing markets, transform old ones, or create new ones" (p.687), while Akaka et al. (2021) support the latter argument. Indeed, explaining market emergence due to societies facing wicked problems, they highlight that although the analysis of macrostructures (as sociocultural systems) could lead to the identification of macro-problems ("institutionalized problems"), changes in micro-foundations (value co-creation practices) and other institutions and institutional arrangements are also needed to cope with such problems.

By observing if and how the solution(s) to solve the problem of the actors and the problem to be solved change, we can understand what kind of change has occurred within the market actual structure and environment (Table 2.1) as detailed in what follows.

Our typology of market systems changes has similarities with other typologies and concepts developed in different disciplines (Table 2.2). For example, in dealing with regime changes in societal systems, Geels and Kemp (2007) identified three basic pathways: reproduction, transformation, and transition. Taylor and Lomax (2022) proposed the Adopt-Adapt-Expand-Respond (AAER) framework to conceptualize systemic change as an adaptive programming tool. In the field of business

Table 2.1 Extent of change the market system, changed "element," and related characteristics

Type of market change (increasing extent of change)	Change of the solution(s) to solve the actors' problem	Change of the actors' problem to solve	Changed market structure	Leveraged property to enact the change
Incremental adjustment	X		Slight revisions to the actual market structure	Elasticity, action on the elements of the context
Transformation	X		Significant revisions to the actual market structure based on the possibilities of the environment	Flexibility, action on the environment
Reforming	X	X	Significant revisions to both the market actual structure and the environment	Plasticity, action on overall solutions and/or problems to solve

Source Authors' elaboration

organization, Bartunek and Moch (1994) introduced three orders of change: incremental change (behavioral adjustments deemed appropriate within the organization), changes in the cognitive frameworks of the organization, and total transcendence of organizational patterns. In dealing with socio-ecological systems, Gunderson and Holling (2002) illustrated the adaptive cycle, which focuses on systems resilience or renewal. In the following sections, we describe the three changes in detail.

First-Order Change: Incremental Adjustment

The incremental adjustment or first-order change is a slight change in the market specific or actual structure based on a better exploitation of its current endowment in terms of the structure of the actors' resources and relationship. This type of adjustment is made possible by elasticity; that is, the possibility of different exploitation of the elements of the actual market. In other words, redundancies and other abilities already embedded in the market structure but not exploited can be leveraged to make minor adjustments (Golinelli 2010). The adjustment can be similar to the "scaling out" of a simple innovation, with changes close to current market solutions, and rapidly becoming routines (Moore et al. 2015) and to resilience (Allenby and Fink 2000; Weick and Sutcliffe 2011), understood as the ability of the system to develop adaptive behaviors capable of reducing the stress arising from contingencies without upsetting its initial conditions. In other words, adjustment is incremental and involves an acceptance of the impact of a possible disruptive event (i.e., a disaster) and a return

Table 2.2 Typologies of systems' change and similar concepts

Types of (market) change	Similar concepts	References
Adjustment	Scaling out "New solution to solve a problem, closely linked to the current regime rapidly integrated as a new routine."	Moore et al. 2015
	Resilience "the capability of a system to maintain its function and despite internal and external changes"	Allenby and Fink 2000; Weick and Sutcliffe 2011
	Reproduction "[..] the system is improved but maintains its basic structure."	Geels and Kemp 2007
	To regain the shape	Donaldson 1987; Pelling et al. 2014
Transformation	"Something uncontrollable, even incomprehensible, a quantum jump."	Watzlawick et al. 1974, p. 38
	New institutional arrangements in place of some others and, potentially, new market infrastructures	Kjellberg et al. 2019
	Transformation "[..] the development trajectories are re-directed through changes in regime rules"	Geels and Kemp 2007
	Change in the contextual and Industrial relations of the system	Best et al. 2012; Erickson and Kuruvilla 1998
	New/transformed value propositions to influence the (re)formation of markets."	Kaartemo et al. 2018
	Transformation as institutional change	Wagener 1993
Reformation	Large System Change "a fundamental revisioning of what is possible and ways of sensemaking that lead to previously unimaginable outcomes"	Waddell et al. 2015
	Transformative systemic change ("scaling up" and "scaling deep")	Jenal 2019
	Transition "a shift to a new socio-technical system and development trajectory"	Geels and Kemp 2007
	A new preferred configuration of the dissipative structure	Gemmill and Smith 1985
	Radical innovation	O'Connor and Ayers 2005; Garcia and Calantone 2002

(continued)

Table 2.2 (continued)

Types of (market) change	Similar concepts	References
	Radical change and disruptive change	Collis 2001; Kezar 2005; Greenwood and Hinings 1996; Nadler and Nadler 1998; Christensen and Overdorf 2000
	New markets result from changing existing conditions: new networks of actors, new resources, and constraints	Sarasvathy and Dew 2005

Source Authors' elaboration

to initial conditions (Donaldson 1987; Pelling et al. 2014) without changing the pre-existing rules nor the orientation of the dominant actors, the key technology, and the knowledge base (Geels and Kemp 2007). An example of adjustment can be found in the case of a retail market challenged by the introduction of a new tax regulation that requires retailers to trace payments and their e-commerce success. A market shaping strategy, informed by a defensive shaping intention (Storbacka and Nenonen 2011), emerged with the goal of transforming the obligation into an opportunity (Megaro 2023). A cash register provider made a simple product update: cash register tracing payments could now also collect other data about customers' purchases and shared online. Thus, a group of already related market actors (including retailers, the cash register provider, meal voucher companies, software developers, and market research companies) envisioned a TO BE market view in which they could share customers' data (a slight change resulting in a new resource in the market structure). In particular, acting as a natural shopping center as opposed to e-commerce, they became able to collect customers' payments through meal vouchers and profile customers to understand preferences, offer discounts, and customize offers.

Second-Order Change: Transformation

The transformation, or second-order change, is a *significant change to the market actual structure* based on a re-selection of components and relationships from among the possibilities of the market's environment. This type of change focuses on flexibility, the possibility of exploiting the various elements of the environment. The involvement of the environment implies a transformation of the relationships of the market with its environment. Such a change has been defined as "second-order change" because, if compared to the "first-order change", it requires the introduction of "something uncontrollable, even incomprehensible, a quantum jump" (Watzlawick et al. 1974, p. 38). The term "transformation" is typically used to describe non-linear changes in systems where the logic coordinating the actions of system actors is re-directed (Geels and Kemp 2007). Although these kinds of change generate unpredictable and riskier outcomes, some also view them as a source of viability when companies lose their ability to succeed within their particular context (Lichtenstein 2000). Revising relationships with the environment also means establishing or adopting some new institutional arrangements in place of others and, potentially, new market infrastructures (Kjellberg et al. 2019). Additionally, it requires assigning meaning to new/transformed value propositions (Kaartemo et al. 2018). Transformation implies a change in the system of relationships (Best et al. 2012) that preserves the network of basic assumptions and principles underlying that system while modifying its form (Erickson and Kuruvilla 1998). To achieve system-wide transformation, all members of the system must communicate and connect to align their efforts to enact change (Harris 2010). According to Wagener (1993), transformation is one of the most complex forms of institutional change that can be successful if other elements of the socioeconomic system adapt or are modified simultaneously. Actors use enactment and sense-making processes to alter their value propositions or develop new ones to influence the (re)formation of markets, thus reducing uncertainty (Kaartemo et al. 2018).

An example of transformation can be drawn from the growing demand for sustainable products raised by the cleaning products market, which requires companies to invest in R&D activities. A market shaping strategy driven by an offensive shaping intention (Storbacka and Nenonen 2011) was undertaken by a group of actors in Italy. A manufacturer of cleaning products, in collaboration with the University of Salerno and the University of Naples Parthenope within the BIOCHELAV project (Polese et al. 2022) is currently attempting to develop and commercialize an innovative sustainable detergent for dishwashers; this product is not only non-polluting but also enables users to re-purpose the used washing machine water as fertilizer. This new solution to the sustainability problem connected to cleaning products requires a significant change in the market. Specifically, the AS IS market structure comprising suppliers and customers (i.e., families and HO.RE.CA users) must be transformed into a TO BE one that foresees the use of detergent as also a fertilizer (a new resource) and includes new relationships with actors not previously included in the market structure, such as designers and managers of public facilities, municipalities, farmers benefiting from grey water reuse, sanitary ware manufacturers, car manufacturers benefiting from circular economy strategies related to cleaning manufacturers, and open innovation agreements with other detergent producers. While collaborators have agreed on the TO BE market structure, collective and individual market shaping strategies are still under evaluation to reduce market multiplicity and define the best approaches for achieving the desired systemic outcomes (e.g., reducing the concurrence among detergent producers). Given the significant change taking place, actors are highly likely to face unexpected consequences.

Third-Order Change: Reformation

Finally, the reformation or third-order change signifies a change in the overall market structure and systemic behavior and subsequently the nature and characteristics of the components of the market structure and the environment, thus coping with a reformation. It relies on plasticity, the possibility of rethinking solutions and the problems themselves, resulting in a new order or organization of the market.

The reformation is a large system change (Waddell et al. 2015); thus it is not a simple revision of what already exists, such as reorganization within existing structures and relationships. Rather, it refers to a simultaneous change in the complex relationships between elements at several levels, a revision of what is possible, and actions that lead to previously unimaginable results. This type of change in the corporate governance field can be referred to as radical change and occurs when all elements of the strategy need to be changed to alter the business strategy model to achieve a new business advantage (Collis 2001). It may also mean revising the core business idea (Golinelli 2010). It can be compared to a radical change, a complex and far-reaching change that requires managers in organizations to gather the necessary capabilities to address it (Christensen and Overdorf 2000) as it involves changes in strategy, structures, work requirements (Nadler and Nadler 1998), and may necessitate a complete overhaul of the organization (Greenwood and Hinings 1996) to create new processes of governance that are more functional and successful (Kezar 2005). A market reformation implies a systemic change that, according to Jenal (2019), occurs when there is a shift in institutions, politics, laws, the rules of the game ("scaling up"), as well as in people's hearts and minds, their values and cultural practices, and the quality of relationships ("scaling deep"). Here, a change in the problem to be solved within the market drives the reorganization of all actors, both in the actual market structure and in the environment. Changing existing conditions can create a new market through a process involving a unique network of actors and implies a commitment to expanding resources and constraints (Sarasvathy and Dew 2005). It causes a transition, a shift to a new trajectory and system that determines changes in the socio-technical system and generates a new network of social groups and a revision of the regime rules (Geels and Kemp 2007). Reformation means the entire system is reorganized in a new preferred configuration of a successful dissipative structure (Gemmill and Smith 1985). Reformation may be compared to radical innovation, understood as a mechanism for rejuvenating organizations. Different organizational structures vary regarding their relationship with R&D and how far down the commercialization path they oversee projects (O'Connor and Ayers 2005). Disruptive innovations create new markets and engulf or destroy existing ones (Garcia and Calantone 2002).

The healthcare system provides an example of reformation in the initiatives undertaken to counter the COVID-19 pandemic, which has imposed new rules and behaviors over the world. To quickly propose solutions to an unexpected problem, a group of actors undertook a market shaping action with a defensive intention. Research institutions, pharmaceutical companies, universities, governments, and healthcare providers shared some TO BE market views aimed at solutions capable of quickly neutralizing the threats caused by the pandemic (market multiplicity reduction). This change led to a significant revisions: actors not previously connected or interacting in different ways (AS IS) had to collaborate to develop vaccines (a new resource), which resulted in warping the traditional research methods, making parallel phases that were previously sequential, and publishing intermediate results (new rules/institutions). Multiple changes occurred at different levels of market, from individual sanitary practices to new vaccine regulations.

Conclusions, Implications, and Future Research Questions

In this manuscript, we assumed the roles of the cartographer and the naturalist (MacInnis 2011) in contributing to conceptualization within the market shaping stream of research.

As *cartographers,* we delineated the importance, the meaning of, and methods for analyzing a market view to be shared among shapers to *reduce the multiplicity* of market views (Kjellberg et al. 2012) for shapers with a common understanding, thus setting the foundation for a much more efficient collaboration and highlighting the interdependence between the AS IS and TO BE market views. Further, we note the importance of considering the inner complexity arising from the enactment of marketing strategies that have unintended consequences and may lead to the emergence of new properties. In particular, in our *detailing* contribution, we bring attention to the possibility of defining a market view based on the market structure and system analysis and sharing an AS IS and TO BE view with collaborators to pursue market shaping strategies effectively. This has important implications, described as follows. First, shapers should envision a TO BE market view incorporating redundancies and alternatives to make their future context elastic/flexible/plastic to further desired changes. Indeed, heterogeneity is a source of adaptation for complex adaptive systems (Rogers et al. 2005). Second, since shapers always want to pursue stabilizing market strategies (Kjellberg et al. 2015) to fully exploit the benefits of their shaping activities, they should consider that excessive institutional constraints of the market would prevent future exploitation of market malleability. Third, the TO BE market structures should be designed in a such a way that they are consonant with (i.e., aligned to) the environment and other market structures to increase their chances of success in the implementation of strategies and synergies with other actors (i.e., resonant behavior) (Barile and Polese 2010). Fourth, since the systemic consequences of these design choices, detectable in the market systems, may differ from the expected ones of the TO BE view given the phenomenon of emergence, shapers should be aware of the complexity, embrace it, and employ their analytical reasoning to draw new AS IS and TO BE views to face such “disturbances”. Finally, practically speaking, it is interesting to note that based on the shared structure of the TO BE market view, shapers could utilize tools (e.g., systems dynamics and agent-based modeling) to design structures and simulate systemic behaviors and try to predict the systemic outcomes of collective shaping. Examples of future research questions related to these implications are reported in Table 2.3.

Playing the role of *naturalists* in contributing to conceptualization within the market shaping stream of research, we leveraged the market view understanding to develop common distinct types of market change (a *differentiating* contribution). The distinction between adjustment, transformation, and reformation, based on the diverse related abilities to change (e.g., elasticity, flexibility, and plasticity) of a market, is beneficial to market shapers since it is based on the extent of change that the market structure should undertake to achieve the desired TO BE market view. This has important implications as described in the following. To begin with, the distinction allows the sharing of the market view to reduce the market multiplicity of collaborative shapers by clearly stating the objectives of a shaping strategy. Second, such a shared "type of change" would be helpful to make a connection between the desired changes and the predictability and management of the consequent systemic outcomes. Indeed, adjustments are easily triggered, and their outcomes seem predictable and manageable since they can be faced with sufficient endowment of actors, resources, and relationships. On the other hand, transformation and reformation are more challenging to manage due to the greater extent of

Table 2.3 Implications and future research related to the market views identification (detailing)

Implications of the market views identification	Future research questions (examples)
i. Shapers' enablement of incorporating redundancies in the TO BE market views to enable future changes ii. Shapers' reflections on avoiding rigidities to chase changes	- What are the methods and capabilities through which shapers could agree on AS IS and TO BE market views? - How much should market views be diffused to avoid possible opportunisms by weakly related collaborators? - What capabilities should shapers show to design malleable structures (able to allow desired further changes with little institutional work for absorbing unexpected shocks)? - How should the best tradeoff be assured between malleability to new situations and stability for current exploitation?
iii. Shapers' design of consonant TO BE market structures	- What are the main elements that should be considered to measure consonance (between market and environment, or among markets) for market shaping? - What about the drivers of consonance between shapers?
iv. Shapers' embracement of systemic complexity	- How should complexity be incorporated into market shaping design and strategy development?
v. Shapers' adoption of tools to model structures and simulate systemic behaviors	- What are the assumptions, procedures, and methods of modeling and simulation?

Source Authors' elaboration

required change, making the related emergent outcomes less predictable. Indeed, in those cases, the outcomes may be emergent, and the elements that give rise to such emergence may be unknown. Thus, the reflections emerging from our typology can be related to causation/predictions and effectuation/control (Sarasvathy 2001) that shapers should be ready to dominate. Indeed, the elasticity to adjustments of the context can addressed by causation/predictions, and changing the solution(s) is manageable with the current market endowment. Flexibility to transformations of the environment and the plasticity of the reformulation of the market are more related to effectuation/control, changing the solution(s) and/or making the problem manageable with a significant change to/reformulation of the market and with more unpredictable outcomes, which shapers should attempt to control more than predict. Furthermore, the proposed typology could be used to share a change type to be pursued by market shapers or other market actors and to label and communicate a change type experienced in a market to reduce market multiplicity due to emergent properties intended as disturbances. Reactions to such disturbances are due to the market system's self-organization. While disturbances manageable by adjustments do not require the actual market structure to change, significant disturbances require a transformation, with changes to the market structure based on the availability of environmental alternatives. Moreover, when the disturbances cannot be accommodated in the environment, a reformation of the system remains the only way to make the market system viable again, the alternative to the lack of stress absorption being collapse. Thus, a new selection among the alternative configurations of actors and resources is made to define a specific structure and order/organization (Polese et al. 2022). Such labeling, in turn, would be beneficial to identify moments in which shaping can be more effective. Indeed, according to Nenonen and Storbacka (2020), since different market directions become possible during unstable periods (identifiable as transformations and reformations), efforts to nudge the market toward a specific direction could be lower. Moreover, during these times, since deeply rooted mental models (Storbacka and Nenonen 2011) are already challenged by new emerging ones, offering alternative ways to overcome instabilities can be seen as a credible promise to regain stability. Finally, with our typology, we contribute to a deeper understanding of the concept of market plasticity (the market's ability to take and retain a form), as highlighted by Nenonen et al. (2014). The market's ability to change, which is related to plasticity, is the extreme expression of a market maintaining viability while reforming and changing solution(s) and/or problems to solve. Examples of future research questions related to these implications are reported in Table 2.4.

Table 2.4 Implications and future research related to the typology of market changes (differentiating)

Implications of the typology of market changes	Future research questions (examples)
1. Shapers sharing the objectives of their strategies	- How can shapers share a market type of change and overcome the problems related to the partial views of markets due to their different goals?
2. Shapers awareness of the extent of change, predictability, and manageability	- What are the specific capabilities through which shapers can undertake and face the outcomes of a pursued adjustment, transformation, or reformulation of a market?
3. Shapers consideration of the dominance of needed causation/predictions or effectuation/control approaches	
4. Market actors labeling and communication of market change type experienced in a market and faced with an escalation logic (from adjustments to reformulation)	- Which are the most effective ways to communicate the change type experienced in the market to shape the market?
5. Shapers support identifying moments in which shaping can be more effective	- How can they identify the moments in which shaping can be more effective?
6. Further understanding of market plasticity	- What are the capabilities that shapers should have to deal with the plasticity of the market that can influence its viability as it reforms, changes solutions, and/or problems to solve?

Source Authors' elaboration

References

Akaka, M.A., Koskela-Huotari, K., Vargo, S.L.: Formalizing service-dominant logic as a general theory of markets: taking stock and moving forward. AMS Rev. 1–15 (2021)

Alderson, W.: Dynamic marketing behavior: a functionalist theory of marketing. Richard D. Irwin, Homewood (1965)

Allenby, B., Fink, J.: Social and ecological resilience: toward inherently secure and resilient societies. Science **24**(3), 347–364 (2000)

Badinelli, R., Barile, S., Ng, I., Polese, F., Saviano, M., Di Nauta, P.: Viable service systems and decision making in service management. J. Serv. Manag. **23**(4), 498–526 (2012)

Baker, J.J., Nenonen, S.: Collaborating to shape markets: Emergent collective market work. Ind. Mark. Manage. **85**, 240–253 (2020)

Barile, S., Saviano, M.: Foundations of systems thinking: the structure-system paradigm. In: AA.VV. (eds.) Contributions to theoretical and practical advances in management—A Viable Systems Approach (VSA). International Printing, Avellino, pp. 1–25 (2011)

Barile, S.: Management sistemico vitale. Parte Prima Decidere in contesti complessi. Giappichelli Editore (2009a)

Barile, S.: The dynamic of informative varieties in the processes of decision making. In: The 3rd International Conference on Knowledge Generation, Communication and Management. Orlando, Florida (2009b)

Barile, S., Polese, F.: Smart service systems and viable service systems: applying systems theory to service science. Serv. Sci. **2**(1/2), 21–40 (2010)

Barile, S., Lusch, R., Reynoso, J., Saviano, M., Spohrer, J.: Systems, networks, and ecosystems in service research. J. Serv. Manag. **27**(4), 652–674 (2016)

Barile, S., Pels, J., Polese, F., Saviano, M.: An introduction to the viable systems approach and its contribution to marketing. J. Bus. Mark. Manag. **5**(2), 54–78 (2012)
Bartunek, J.M., Moch, M.K.: Third-order organizational change and the western mystical tradition. J. Organ. Chang. Manag. **7**(1), 24–41 (1994)
Beer, S.: The Heart of Enterprise. Wiley, Suffolk, UK (1979)
Beer, S.: Diagnosing the System for Organizations. Wiley, Suffolk, UK (1985)
Beninger, S., Francis, J.N.: Collective market shaping by competitors and its contribution to market resilience. J. Bus. Res. **122**, 293–303 (2021)
Best, A., Greenhalgh, T., Lewis, S., Saul, J.E., Carroll, S., Bitz, J.: Large-system transformation in health care: a realist review. Milbank Q. **90**(3), 421–456 (2012)
Brodie, R.J., Löbler, H., Fehrer, J.A.: Evolution of service-dominant logic: towards a paradigm and metatheory of the market and value cocreation? Ind. Mark. Manage. **79**, 3–12 (2018)
Capra, F.: The web of life. Harper Collins Publishers, London (1996)
Chandler, J.D., Vargo, S.L.: Contextualization and value-in-context: how context frames exchange. Mark. Theory **11**(1), 35–49 (2011)
Checkland, P.: Systems Thinking, Systems Practice (30 Year Retrospective). Wiley, Oxford UK (1999)
Checkland, P.: Systems Thinking, Systems Practice. John Wiley and Sons Ltd., Chichester (1997)
Christensen, C.M., Overdorf, M.: Meeting the challenge of disruptive change. Harv. Bus. Rev. **78**(2), 66–77 (2000)
Collis, D.: When industries change: the future of higher education. Contin. High. Educ. Rev. **65**, 7–24 (2001)
Diaz Ruiz, C.A., Baker, J.J., Mason, K., Tierney, K.: Market-scanning and market-shaping: why are firms blindsided by market-shaping acts? J. Bus. Ind. Mark. **35**(9), 1389–1401 (2020)
Dixon, D.F.: A social systems approach to marketing. Soc. Sci. Q. **48**, 164–173 (1967)
Domegan, C., McHugh, P., Devaney, M., Duane, S., Hogan, M., Broome, B.J., Piwowarczyk, J.: Systems-thinking social marketing: conceptual extensions and empirical investigations. J. Mark. Manag. **32**(11–12), 1123–1144 (2016)
Donaldson, L.: Strategy and structural adjustment to regain fit and performance: in defence of contingency theory. J. Manage. Stud. **24**(1), 1–24 (1987)
Elg, U., Deligonul, S.Z., Ghauri, P.N., Danis, W., Tarnovskaya, V.: Market-driving strategy implementation through global supplier relationships. Ind. Mark. Manage. **41**, 919–928 (2012)
Erickson, C.L., Kuruvilla, S.: Industrial relations system transformation. ILR Rev. **52**(1), 3–21 (1998)
Fehrer, J.A., Conduit, J., Plewa, C., Li, L.P., Jaakkola, E., Alexander, M.: Market shaping dynamics: interplay of actor engagement and institutional work. J. Bus. Ind. Mark. **35**(9), 1425–1439 (2020)
Fisk, G.: Marketing Systems. Harper and Row, New York (1967)
Flaig, A., Kindström, D., Ottosson, M.: Market-shaping phases—a qualitative meta-analysis and conceptual framework. AMS Rev. **11**(3–4), 354–374 (2021)
Fligstein, N.: The Architecture of Markets. Princeton University Press (2001)
Garcia, R., Calantone, R.: A critical look at technological innovation typology and innovativeness terminology: a literature review. J. Prod. Innov. Manag. Int. Publ. Prod. Dev. Manag. Assoc. **19**(2), 110–132 (2002)
Geels, F.W., Kemp, R.: Dynamics in socio-technical systems: typology of change processes and contrasting case studies. Technol. Soc. **29**(4), 441–455 (2007)
Gemmill, G., Smith, C.: A dissipative structure model of organization transformation. Hum. Relat. **38**(8), 751–766 (1985)
Giesler, M., Fischer, E.: Market system dynamics. Mark. Theory **17**(1), 3–8 (2017)
Golinelli, G.M., Barile, S., Saviano, M., Polese, F.: Perspective shifts in marketing: toward a paradigm change? Serv. Sci. **4**(2), 121–134 (2012)
Golinelli, G.M. (2010). *Viable Systems Approach (VSA). Governing Business Dynamics*, Cedam: Italy.

Gosling, M., Richard, J., Seo, Y.: Markets and market boundaries: a social practice approach. J. Serv. Theory Pract. **27**(2), 408–426 (2017)
Greenwood, R., Hinings, C.R.: Understanding radical organizational change: bringing together the old and the new institutionalism. Acad. Manag. Rev. **21**(4), 1022–1054 (1996)
Gunderson, L.H., Holling, C.S. (eds.): Panarchy: understanding transformations in human and natural systems. Island press (2002)
Harris, A.: Leading system transformation. Sch. Leadersh. Manag. **30**(3), 197–207 (2010)
Hauser, J., Tellis, G.J., Griffin, A.: Research on innovation: a review and agenda for marketing science. Mark. Sci. **25**(6), 687–717 (2006)
Hawa, J., Baker, J., Plewa, C.: Composing markets: a framework of intentionality in market-shaping. J. Bus. Res. **121**, 47–57 (2020)
Holbrook, M.B.: Adventures in complexity: an essay on dynamic open complex adaptive systems, butterfly effects, self-organizing order, coevolution, the ecological perspective, fitness landscapes, market spaces, emergent beauty at the edge of chaos, and all that jazz. Acad. Mark. Sci. Rev. **6**(1), 1–184 (2003)
Jaakkola, E.: Designing conceptual articles: four approaches. AMS Rev **10**, 18–26 (2020)
Jaworski, B., Kohli, A.K., Sahay, A.: Market-driven versus driving markets. J. Acad. Mark. Sci. **28**(1), 45–54 (2000)
Jenal, M.: Let's put systemic change to rest (2019)
Kaartemo, V., Kowalkowski, C., Edvardsson, B.: Enhancing the understanding of processes and outcomes of innovation: the contribution of effectuation to SD logic. In: Sage handbook of service-dominant logic, pp 522–535 (2018)
Kaartemo, V., Nenonen, S., Windahl, C.: Institutional work by market-shaping public actors. J. Serv. Theory Pract. **30**(4/5), 401–435 (2020)
Kezar, A.: Consequences of radical change in governance: a grounded theory approach. J. High. Educ. **76**(6), 634–668 (2005)
Kim, J.: Making sense of emergence. Philos. Stud. **95**, 3–36 (1999)
Kjellberg, H., Helgesson, C.F.: Multiple versions of markets: multiplicity and performativity in market practice. Ind. Mark. Manage. **35**(7), 839–855 (2006)
Kjellberg, H., Murto, R.: Theorizing Markets. AMS Rev. **11**(3–4), 207–215 (2021)
Kjellberg, H., Azimont, F., Reid, E.: Market innovation processes: balancing stability and change. Ind. Mark. Manage. **44**, 4–12 (2015)
Kjellberg, H., Hagberg, J., Cochoy, F.: Thinking market infrastructure: barcode scanning in the US grocery retail sector, 1967–2010. In: Kornberger, M., Bowker, G., Elyachar, J., Mennicken, A., Miller, P., Nucho, J. and Pollock, N. (eds.): Thinking Infrastructures (Research in the Sociology of Organizations, vol. 62, pp. 207–232. Emerald Publishing Limited (2019)
Kjellberg, H., Storbacka, K., Akaka, M., Chandler, J., Finch, J., Lindeman, S., Löbler, H., Mason, K., McColl-Kennedy, J., Nenonen, S.: Market futures/future markets: research directions in the study of markets. Mark. Theory **12**(2), 219–223 (2012)
Kleinaltenkamp, M., Conduit, J., Plewa, C., Karpen, I.O., Jaakkola, E.: Engagement-driven institutionalization in market shaping: synchronizing and stabilizing collective engagement. Ind. Mark. Manage. **99**, 69–78 (2021)
Kullak, F.S., Fehrer, J.A., Baker, J.J., Woratschek, H., Sam-Cobbah, J.: Shaping market systems for social change in emerging economies. Ind. Mark. Manage. **100**, 19–35 (2022)
Kumar, N., Scheer, L., Kotler, P.: From market driven to market driving. Eur. Manag. J. **18**, 129–142 (2000)
Layton, R., Duffy, S.: Path dependency in marketing systems: where history matters and the future casts a shadow. J. Macromark. **38**(4), 400–414 (2018)
Layton, R.A.: Marketing systems—a core macromarketing concept. J. Macromark. **27**(3), 227–242 (2007)
Layton, R.A.: Towards a theory of marketing systems. Eur. J. Mark. (2011)
Layton, R.A.: Formation, growth, and adaptive change in marketing systems. J. Macromark. **35**(3), 302–319 (2014)

Lewis, R.J., Erickson, L.G.: Marketing functions and marketing systems: a synthesis. J. Mark. **33**(3), 10–14 (1969)
Lichtenstein, B.B., Uhl-Bien, M., Marion, R., Seers, A., Orton, J.D., Schreiber, C.: Complexity leadership theory: an interactive perspective on leading in complex adaptive systems (2006)
Lichtenstein, B.B.: Self-organized transitions: a pattern amid the chaos of transformative change. Acad. Manag. Perspect. **14**(4), 128–141 (2000)
Lukka, K., Vinnari, E.: Domain theory and method theory in management accounting research. Account. Audit. Account. J. **27**(8), 1308–1338 (2014)
Lusch, R.F., Vargo, S.L.: Service-Dominant Logic: Premises. Possibilities, Cambridge University Press, New York, NY, Perspectives (2014)
MacInnis, D.J.: A framework for conceptual contributions in marketing. J. Mark. **75**(4), 136–154 (2011)
Markose, S.M.: Computability and evolutionary complexity: markets as complex adaptive systems (CAS). Econ. J. **115**, 159–192 (2005)
Megaro, A.: Smart cash point in a B2B market to enable service innovation: towards transparency and shared intentions: a case study. J. Ind. Bus. Ind. Mark. (2023)
Mele, C., Pels, J., Polese, F.: A brief review of systems theories and their managerial applications. Serv. Sci. **2**(1–2), 126–135 (2010)
Mele, C., Pels, J., Storbacka, K.: A holistic market conceptualization. J. Acad. Mark. Sci. **43**, 100–114 (2015)
Mele, C., Russo-Spena, T., Kaartemo, V.: The impact of coronavirus on business: developing service research agenda for a post-coronavirus world. J. Serv. Theory Pract. **31**(2), 184–202 (2021)
Mengist, W., Soromessa, T., Legese, G.: Method for conducting systematic literature review and meta-analysis for environmental science research. MethodsX **7**, 100777 (2020)
Mitleton-Kelly, E.: Ten principles of complexity and enabling infrastructures. In: Mitleton-Kelly, E. (ed.) Complex Systems and Evolutionary Perspectives on Organisations, pp. 23–50. Elsevier Science (2003)
Möller, K., Nenonen, S., Storbacka, K.: Networks, ecosystems, fields, market systems? Making sense of the business environment. Ind. Mark. Manage. **90**, 380–399 (2020)
Moore, M.L., Riddell, D., Vocisano, D.: Scaling out, scaling up, scaling deep: strategies of non-profits in advancing systemic social innovation. J. Corp. Citizsh. **58**, 67–84 (2015)
Nadler, D.A., Nadler, M.B.: Champions of change: How CEOs and their companies are mastering the skills of radical change. Jossey-Bass (1998)
Nenonen, S., Storbacka, K.: Don't adapt, shape! use the crisis to shape your minimum viable system—and the wider market. Ind. Mark. Manage. **88**, 265–271 (2020)
Nenonen, S., Storbacka, K.: Market-shaping: navigating multiple theoretical perspectives. AMS Rev **11**, 336–353 (2021)
Nenonen, S., Kjellberg, H., Pels, J., Cheung, L., Lindeman, S., Mele, C., Sajtos, L., Storbacka, K.: A new perspective on market dynamics: market plasticity and the stability–fluidity dialectics. Mark. Theory **14**(3), 269–289 (2014)
Nenonen, S., Storbacka, K., Frethey-Bentham, C.: Is your industrial marketing work working? Developing a composite index of market change. Ind. Mark. Manage. **80**, 251–265 (2019)
Nenonen, S., Storbacka, K., Sklyar, A., Frow, P., Payne, A.: Value propositions as market-shaping devices: a qualitative comparative analysis. Ind. Mark. Manage. **87**, 276–290 (2020)
Nightingale, A.: A guide to systematic literature reviews. Surg. Infect. (Larchmt.) **27**(9), 381–384 (2009)
Nilsson, J., Murto, R., Kjellberg, H.: Influencer marketing and the 'gifted' product: framing practices and market shaping. J. Mark. Manag. **39**(11–12), 982–1011 (2023)
Nyström, A.G., Kaartemo, V.: Developing Delphi methodology for studying future market change. J. Bus. Ind. Mark. **37**(13), 124–141 (2022)
O'Connor, G.C., Ayers, A.D.: Building a radical innovation competency. Res. Technol. Manag. **48**(1), 23–31 (2005)

Ozdamar Ertekin, Z., Atik, D., Murray, J.B.: The logic of sustainability: Institutional transformation towards a new culture of fashion. J. Mark. Manag. **36**(15–16), 1447–1480 (2020)

Patvardhan, S., Ramachandran, J.: Shaping the future: strategy making as artificial evolution. Organ. Sci. **31**(3), 671–697 (2020)

Pedeliento, G., Mangiò, F., Murtas, G., Andreini, D.: Market system dynamics (MSD): a process-oriented review of the literature. AMS Review **13**, 173–195 (2023)

Pelling, M., O'Brien, K., Matyas, D.: Adaptation and transformation, climatic change. Adv. Clim. Chang. Adapt. Risk Manag. **10** (2014)

Pels, J., Barile, S., Saviano, M., Polese, F., Carrubbo, L.: The contribution of VSA and SDL perspectives to strategic thinking in emerging economies. Manag. Serv. Qual. (2014)

Pels, J., Mele, C., Spano, M.: From market driving to market shaping: impact of a language shift. J. Bus. Ind. Mark. **38**(1), 155–169 (2023)

Peters, L.D., Nenonen, S., Polese, F., Frow, P., Payne, A.: Viability mechanisms in market systems: prerequisites for market shaping. J. Bus. Ind. Mark. **35**(9), 1403–1412 (2020)

Polese, F., Pels, J., Tronvoll, B., Bruni, R., Carrubbo, L.: A4A relationships. J. Serv. Theory Pract. **27**(5), 1040–1056 (2017)

Polese, F., Sarno, D., Carrubbo, L., Megaro, A.: Sustainable marketing transitions: insights from service-dominant logic. In: Proceedings of SIM Conference, pp. 20–21. Salerno, Italy (2022)

Rogers, E.M., Medina, U.E., Rivera, M.A., Wiley, C.J.: Complex adaptive systems and the diffusion of innovations. Innov. J. Public Sect. Innov. J. **10**(3) (2005)

Sarasvathy, S.D.: Causation and effectuation: toward a theoretical shift from economic inevitability to entrepreneurial contingency. Acad. Manag. Rev. **26**(2), 243–263 (2001)

Sarasvathy, S.D., Dew, N.: New market creation through transformation. J. Evol. Econ. **15**(5), 533–565 (2005)

Senge, P.: Comments on Illuminating the blind spot: leadership in the context of emerging worlds. http://www.dialogonleadership.org/SengeComment.html (2001)

Storbacka, K., Nenonen, S.: Markets as configurations. Eur. J. Mark. **45**(1/2), 241–258 (2011)

Struben, J., Lee, B.H., Bingham, C.B.: Collective action problems and resource allocation during market formation. Strat. Sci. **5**(3), 245–270 (2020)

Taillard, M., Peters, L.D., Pels, J., Mele, C.: The role of shared intentions in the emergence of service ecosystems. J. Bus. Res. **69**(8), 2972–2980 (2016)

Taleb, N.N.: The black swan: the impact of the highly improbable. Penguin Books, London (2008)

Taylor, B., Lomax, J.: Systems, sapiens, and systemic change in markets: the adopt-adapt-expand-respond framework. In: Ton, G., Vellema, S. (eds.): Theory-Based Evaluation of Inclusive Business Programmes. IDS Bulletin 53.1, Brighton: IDS (2022)

Tronvoll, B., Barile, S., Caputo, F.: A systems approach to understanding the philosophical foundation of marketing studies. Soc. Dyn. Syst. Perspect. 1–18 (2018)

Vargo, S.L., Peters, L., Kjellberg, H., Koskela-Huotari, K., Nenonen, S., Polese, F., Sarno, D., Vaughan, C.: Emergence in marketing: an institutional and ecosystem framework. J. Acad. Mark. Sci. **51**(1), 2–22 (2023)

Vargo, S.L., Lusch, R.F.: Inversions of Service-Dominant Logic. Mark. Theory **14**(3), 239–248 (2014)

Vargo, S.L., Lusch, R.F. (eds.): The SAGE Handbook of Service-Dominant Logic. SAGE Publications Inc., London, UK (2019)

Vargo, S.L., Koskela-Huotari, K., Baron, S., Edvardsson, B., Reynoso, J., Colurcio, M.: A systems perspective on markets–toward a research agenda. J. Bus. Res. **79**, 260–268 (2017)

Vargo, S.L., Wieland, H., Akaka, M.A.: Innovation through institutionalization: a service ecosystems perspective. Ind. Mark. Manage. **44**, 63–72 (2015)

von Glasersfeld, E.: The radical constructivist view of science. Found. Sci. **6**(1–3), 31–43 (2001)

Waddell, S., Waddock, S., Cornell, S., Dentoni, D., McLachlan, M., Meszoely, G.: Large systems change: an emerging field of transformation and transitions. J. Corp. Citizsh. **58**, 5–30 (2015)

Wagener, H.J. (1993). Some theory of systemic change and transformation. *On the theory and policy of systemic change*, 1–20.

Watzlawick, P., Weakland, J.H., Fisch, R.: Change: Principles of Problem Formation and Problem Resolution. WW.Nortonand Company, NY (1974)
Weick, K.E.: Sensemaking in organizations (3). Sage (1995)
Weick, K.E., Sutcliffe, K.M.: Managing the unexpected: Resilient performance in an age of uncertainty (8). John Wiley and Sons (2011)
Wieland, H., Nariswari, A., Akaka, M.A.: On managerial relevance: reconciling the academic-practitioner divide through market theorizing. AMS Rev. **11**, 252–271 (2021)
Wieland, H., Polese, F., Vargo, S.L., Lusch, R.F.: Toward a service (eco) systems perspective on value creation. Int. J. Serv. Sci. Manag. Eng. Technol. (IJSSMET) **3**(3), 12–25 (2012)
Windahl, C., Karpen, I.O., Wright, M.R.: Strategic design: orchestrating and leveraging market-shaping capabilities. J. Bus. Ind. Mark. **35**(9), 1413–1424 (2020)
Zietsma, C., McKnight, B.: Building the iron cage: institutional creation work in the context of. In: Institutional work: Actors and agency in institutional studies of organizations, vol. 143 (2009)

Chapter 3
Complexity, Frictions, and Evolution in Service Ecosystems

Alessia Anzivino, Debora Sarno, and Roberta Sebastiani

Abstract This conceptual work focuses on service ecosystems dynamics, since they can be crucial for innovation, resilience, and sustainability. Specifically, in this chapter these dynamics have been considered in relation to actors' perspectives and perceptions on service ecosystem complexity and frictions. Recognizing the importance of such elements in the evolution of service ecosystems, it offers a framework for matching the three service ecosystem complexity states (ordered, manageable, and free complexity), the most common frictions (dilemmas, trade-offs, and paradoxical tensions), and the main systems dynamics drivers (emergence and institutionalization). It also provides illustrations in the realm of healthcare systems and their relations with digitalization and new technologies. The study offers labels for contexts and supports the selection of strategies to cope with complexity and related frictions to both academics and practitioners.

Introduction

The service ecosystem perspective from service-dominant logic (S-D L, Vargo and Lusch 2016) enables the understanding of actors' interactions for value co-creation not just considering individual needs and expectations, but framing them in the wider context in which actors operate. The characteristics of service ecosystems and the related dynamics, thus, are particularly relevant to make sense of actors' actions and reactions, and eventually analyze and envision strategies to foster the evolution of

A. Anzivino (✉) · R. Sebastiani
Università Cattolica del Sacro Cuore, Milan, Italy
e-mail: alessia.anzivino@unicatt.it

R. Sebastiani
e-mail: roberta.sebastiani@unicatt.it

D. Sarno
University of Naples Parthenope, Naples, Italy
e-mail: debora.sarno@uniparthenope.it

D. Sarno et al. (eds.), *Complexity and Emergence in Market Ecosystems*, New Economic Windows, https://doi.org/10.1007/978-3-031-81942-1_3

service ecosystems, for example for targeting new business opportunities, sustainable development, or sustainability transitions to cope with grand challenges.

We particularly deepen complexity and frictions because complexity is the leit motif of service ecosystems (Gummesson et al. 2019). It arises from actors' interactions, due to inherent multidimensionality and heterogeneity of actors and generates a series of frictions.

We leverage the institutional view of S-D logic, according to which institutional arrangements (including rules, social norms, and a shared language) play a central role in facilitating interactions by bringing individuals together and providing a shared framework that guides behavior and communication (Akaka and Chandler 2019), helping actors to interpret and navigate their environment (Cheung and McColl-Kennedy 2015; Edvardsson et al. 2014; Thornton et al. 2012). Through learning and interaction, actors are socialized into these institutional logics, which are often specific to particular contexts or organizational fields. Each organizational field operates within its own set of institutional logics, creating complex contextual frameworks that actors must understand and maneuver through in their interactions.

We notice that institutional arrangements, especially worldviews and practices of actors within service ecosystems, are not always aligned (Gallan et al. 2019). Such incompatible prescriptions for actions deriving from institutional complexity (Greenwood et al. 2011) brings to frictions and tensions (Banoun et al. 2016; Vargo and Lusch 2011; Thornton and Ocasio 1999; McColl-Kennedy et al. 2020). Given this, such frictions should be addressed to allow institutional arrangements to co-evolve and adjust over time (Cherrier 2024) and could also be sources of creativity and dynamism (Tòth et al. 2022).

We are interested in deepening our understanding of the service ecosystem dynamics, and particularly the service ecosystem self-adjustment (i.e. changes in service ecosystems, Vargo and Lusch 2016). We relate these dynamics to actors' perceptions of service ecosystem complexity and related frictions. In psarticular, we provide a typology of service ecosystem complexity states according to actors' perceptions, relating them to the most frequent frictions identified in the literature. This can provide labels for contexts and support in the selection of strategies to cope with complexity and related frictions to both academics and practitioners.

In the following, we deepen the concept of complexity (Sect. 3.2) and frictions (Sect. 3.3) to look for a framework able to integrate them within service ecosystem dynamics (Sect. 3.4). Conclusions follow.

The Duality of Complexity: A Source of Uncertainty and a Driver for the Evolution of Service Ecosystems

The word complexity is not easy to tackle, and the myriad of studies that have been carried on it resulted in a plethora of definitions (Barile et al. 2018). In S-D logic, editorials have been published on the topic (Gummesson et al. 2019), and the term has been mostly mentioned in connection to complex adaptive systems, thus mainly referring to the non-linearity of the relationships among interacting elements and the related emergent outcomes (Vargo et al. 2020).

Drawing on systems theory, and specially on the Viable Systems Approach (Barile 2024; Golinelli 2010), moreover, complexity has been referred to service systems as a constituent part of service ecosystems (Brozovic and Tregua 2021), related to survival and viability (Barile et al. 2016). In particular, since service systems have the primary purpose of survival (Beer 1989; Golinelli 2010), they pursue it by adapting to the changing environment, creating conditions for viability. Complexity, then, comes into the game due to interactions and the emergence of new systems' properties, and changes the context again.

Thus, there is not only an objective and quantitative view on complexity, which is related to the number and variety of actors and resources and relationships among them. To the opposite, it has been added to such objective view on complexity, a subjective and qualitative one (Barile and Saviano 2018). Thus, complexity is also a condition that arises when actors "fail to understand the emerging, unfamiliar dynamics" (Barile et al. 2016, p. 657), shifting the focus from object to subject (i.e., the actor, with its perspective). In other words, something complex may be complicated or certain dependings on the knowledge endowment (Gaeta et al. 2019) and capabilities (Teece and Pisano 2003) of the actor that observes/experiences it (Badinelli et al. 2012). The complexity of the decisions may also derive from the misalignment of the goals and perspectives of actors (Reynoso et al. 2018). However, as complexity theory widely recognised, such a multiplicity of perspectives and knowledge is also one of the main sources for addressing complexity (Desai 2010).

Over time, authors have also attributed the adjective "complex" to resource integration patterns (Sklyar et al. 2019), to decisions to make (Badinelli et al. 2012), to the service ecosystems in general (Vink et al. 2021) mainly due to emergent properties that manifest due to actors interactions and may become institutionalized or not in the service ecosystem (Vargo et al. 2023). Furthermore, the above mentioned institutional view of S-D logic is also related to the concept of "institutional complexity" (Greenwood et al. 2011), which is the "multiplicity of institutional arrangements in service ecosystems, which confronts actors with incompatible prescriptions for action" (Siltaloppi et al. 2016). Since institutional arrangements are related to how actors make sense of their reality and act according to it (Vargo and Lusch 2017), institutional complexity can result in "intersecting and overlapping institutional arrangements that can impose diverging views on what value is" (Siltaloppi et al. 2016). Thus, it can foster conflicting interests among actors, and incompatible ways for integrating resources to cocreate value". However, "the degree of incompatibility among institutional arrangements, and actors' embeddedness within them, shapes how actors experience and respond to institutional complexity" (Siltaloppi et al. 2016). Thus, institutional complexity can drive change and innovation due to the possible emergence of novel solutions.

In summary, this review of the literature shows that complexity in service ecosystems is contextual and actor-dependent. We can distinguish among complexity that is internal to actors and organizations versus complexity of the service ecosystems;

complexity in operations versus complexity in governance and decision making; and quantitative/structural versus qualitative/behavioural complexity. However, what is much more interesting is the duality of complexity in service ecosystems (see Fig. 3.1): complexity is a source of uncertainty (due to misalignment of perspectives, lack of understanding, impossibility to predict emergent properties, etc.), preventing recurrent patterns of resource integration and value co-creation, but it is also a driver for change, soliciting actors to face it, find new solutions, and thus fostering the evolution of service ecosystems. These are the reasons why complexity is so important to service ecosystem perspective.

This conclusion is in line and further explains the evolution of service ecosystems due to complexity informed by complexity science. In another chapter of this book (Sarno et al. 2025a), indeed, with the aim of further understanding radical systems changes in S-D logic thanks to complexity science, the dynamics of service ecosystem have been presented as evolving in a "region of emergent complexity delimitated by order and chaos". In particular, order means organization and stability, while chaos means lack of any organization, instability/untenability. However, too much order can imply rigidity and prevent further change, while chaos could be a source of creativity and transformation. This is why complexity, with its dual role, is fundamental to service ecosystem viability, and why it should be kept near the edge of chaos, where such a viability is the maximum (Sarno et al., 2025a). Focusing on an institutional view of the service ecosystem states (as least as perceived by actors), we can conclude that (see Fig. 3.2):

- In the *order state*, institutional arrangements (norms, practices, etc.) stably enable and constrain value co-creation (even if not in the best possible way—see the example of the "QWERTY" keyboard (Arthur 1994);
- The *complexity state* is characterized by different degrees of institutional complexity (Siltaloppi et al. 2016);

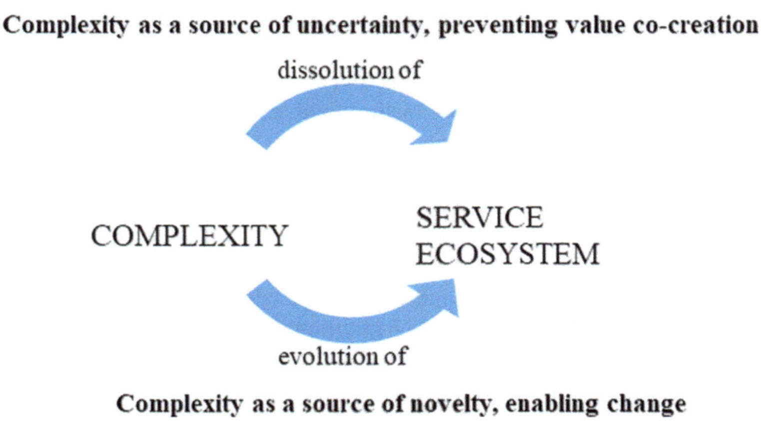

Fig. 3.1 The duality of complexity in service ecosystems dynamics elaboration. *Source* The authors'

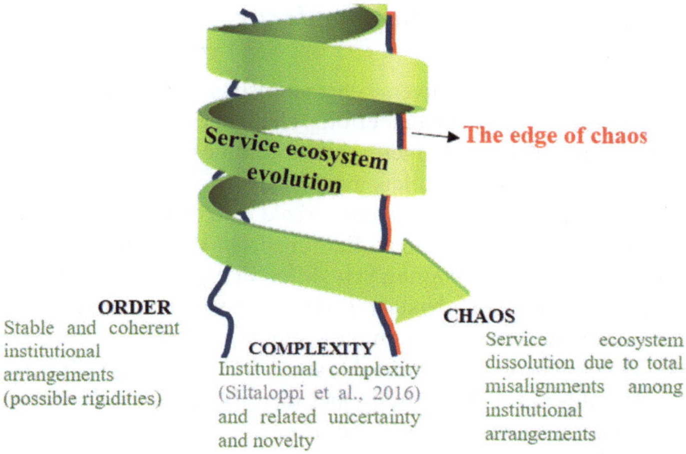

Fig. 3.2 Service ecosystem states and service ecosystems evolution according to an institutional view. Elaborated from Sarno et al. 2025a)

- In the *chaos state*, the service ecosystem may dissolve due to total misalignment among institutional arrangements and, consequently, actors perceptions and behaviors.

Addressing Frictions in Complexity

The duality of complexity in service ecosystem requires actors willing to survive to face and embrace uncertainty and exploit novelties (Vargo et al. 2023).

Nevertheless, worldviews and practices of actors within service ecosystems are not always aligned (Gallan et al. 2019): when there are mismatches in the perspectives and approaches of these actors, frictions and tensions are likely to emerge (Banoun et al. 2016; Vargo and Lusch 2011).

Since the early 1980s, scholars have focused on frictions in the form of paradox, dialectics, and dualities, and have considered tensions as interrelating threads that continuously shape and inform each other throughout time (Smith et al. 2017). More than 30 years ago, some scholars (Poole and Van de Ven 1989) discussed the role of tensions in business and emphasised the function of the resulting dynamics when dealing with exploration and exploitation.

Different Types of Frictions in Complexity

When heterogenous actors with diverse characteristics, experiences, and objectives attempt to collaborate and/or pool their efforts, challenges often emerge. This heterogeneity, as also mentioned before, increases the complexity of the value co-creation processes, making frictions or conflicts more likely.

Also frictions, as complexity, are characterized by a duality and have had several and sometimes contrasting definitions. Starting from the main definitions, in this section we are interested in explaining how different types of frictions are related to complexity in order to better understand service ecosystem dynamics.

Many terms are used to study complex situations (such as sustainability and sustainability development and transition): some studies distinguish between paradoxes, tensions, and trade-offs to study and identify frictions (Haffar and Searcy 2017). In particular, tensions between two or more criteria could lead to trade-offs or synergies. In the presence of resource constraints, tensions involve competition and lead to trade-offs; when conditions coexist, they are referred to as a paradox. Other studies (e.g. Putman et al. 2016) use other terms such as tensions, dualism, duality, dialectics, contradictions, and paradoxes. Tensions refer to the anxiety, stress and other negative factors when making decisions and moving forward in different organisational settings. Dualism considers the presence of opposite poles that can create tensions but can also be separated; duality refers to the interdependence of opposites that are linked in a both/and relationship where they coexist without being mutually exclusive or antagonistic. Dialectics emphasizes the unity of opposites and the forces or processes that unite them. Contradictions consider bipolar opposites that are mutually exclusive yet interdependent, defining and potentially negating each other. Paradoxes are defined as "contradictions that persist over time, impose and reflect on each other, and develop into seemingly irrational or absurd situations because their continuity creates situations in which options appear mutually exclusive, making it difficult to choose between them" (Putman et al. 2016, pp. 8).

For the purpose of this chapter (before making the reader lost among the diverse viewpoints and definitions), we follow Wannags and Gold (2020) who classify frictions as: trade-off, dilemma, tension, and paradox. This is in line with the most commonly discussed terms in sustainable management research (Hahn et al. 2018, 2015, 2010). Specifically, we will consider different types of frictions: dilemma, trade-off, and paradoxical tensions (that include tensions and paradoxes), as follows (Fig. 3.3):

- Trade-off refers to situations where it is impossible find and optimal choice, and a sacrifice in an area is required to gain a benefit in another area (Wannags and Gold 2020; Haffar and Searcy 2017).
- Dilemma refers to conditions where an action is required but the choice involves the gain of one benefit and the loss of another benefit of equal value (Wannags and Gold 2020; Olsen et al. 2007).

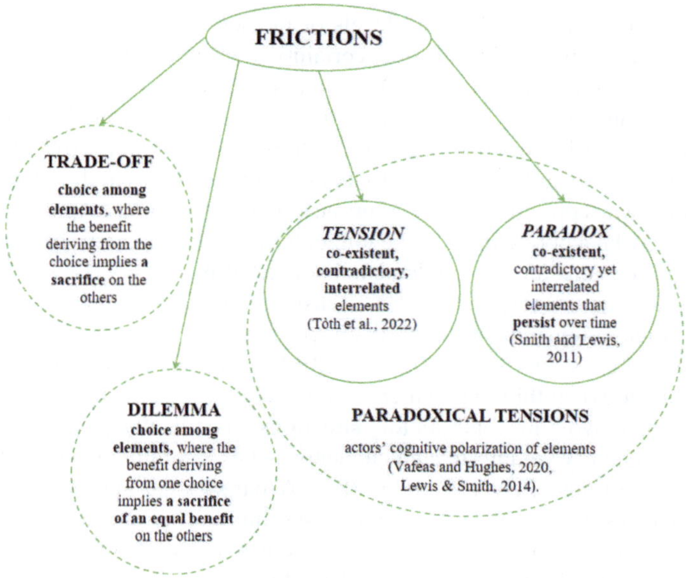

Fig. 3.3 Frictions—dilemmas, trade-off and paradoxical tensions—in complexity elaboration. *Source* The authors'

- Tension refers to co-existent, contradictory, interrelated elements (Tòth et al. 2022). They are also understood as the paradoxical relationships between the opposite poles of a paradox (Hahn et al. 2018).
- Paradox is characterized by two opposing (opposite or contradictory) propositions that are incompatible when taken together, but consistent when taken singularly. Furthermore, paradoxes persist over time and evolve into a situation that may be absurd or irrational (Wannags and Gold 2020; Pole and Van de Ven 1989; Putman et al. 2016).
- Paradoxical tensions are related to actors' cognitive polarization: when actors isolate elements, disregarding or concealing their interdependence, tensions are cognitively and socially framed as paradoxical (Vafeas and Hughes 2020).

Given these definitions, dilemmas and trade-offs are situations in complex contexts, in which the choice of one alternative neglects completely or partially the other and this selection does not provide creative or dynamic solutions (Pole and Van de Ven 1989; Harari 2015).

Tensions, widely discussed in organizational literature, have been described as 'the engine [of the human species] responsible for creativity and dynamism' (Harari 2015, pp. 183–184). They are also 'coexisting, contradictory, interrelated differences within and between organisations that reflect conflicting, non-combinable viewpoints or intentions' (Tòth et al. 2022, pp. 438). With this in mind, it is clear that identifying and managing tensions can lead to innovative and disruptive solutions. Tensions can manifest themselves at different levels of the service ecosystem and generate emotions such as anxiety, discomfort, uncertainty or even paralysis in those who are confronted with them (Putnam et al. 2016). In complex contexts, as organisations or inter-organisational networks change, new forces emerge to reconcile stability with change, creating tensions for actors faced with these contradictions (Tòth et al. 2022). Tensions need to be handled with sensitivity as actions are often high-risk, and trade-offs may not always provide the best solution. Compromising on too many decisions that matter may be a poor choice in the longer term (Boltanski and Thévenot 2006).

Moreover, tensions are the fundamental source of paradoxes (Lewis 2000) that arise from how actors respond to these tensions, and consequently when actors polarise elements, tensions are cognitively perceived as paradoxical (Vafeas and Hughes 2020; Lewis and Smith 2014). In 2011, Smith and Lewis focused on the concept of paradoxical thinking, which is considered more effective than defensive strategies such as avoiding contradictory situations. Specifically, paradoxes could be seen as 'contradictory yet interrelated elements that exist simultaneously and persist over time' (Smith and Lewis 2011, p. 382). A paradoxical approach to tensions 'assumes that tensions are integral to complex systems and that sustainability depends on attending to contradictory yet interwoven demands simultaneously' (Smith and Lewis 2011, p. 397). Paradoxes are thus characterised by their underlying tensions, and actors' responses take tensions into account simultaneously (Smania et al., 2024; Smith and Lewis, 2011). Consequently, paradoxes support a dynamic equilibrium, and tensions and contradictions could be the basis for improved performance if effectively managed (Smania et al. 2024; Tòth et al. 2022). Smith and Lewis (2011) have identified four typologies of paradoxical tensions: belonging, performing, organising and learning. The first—the belonging paradox—relates to complexity and plurality; tensions arise between individuals and groups as they seek homogeneity and distinction. This paradox refers to tensions in sustainability that involve heterogeneous roles and values and could be overcome by considering a unified approach to sustainability that aligns personal values and identities with organisational ones (Anzivino et al. 2021; Smith and Lewis 2011; Sundaramurthy and Lewis 2003). Performing paradox refers to tensions arising from the heterogeneity of actors and stakeholders in organisations, and these tensions often manifest as competing strategies and goals and imply the need to achieve these strategies and goals (Ozanne et al. 2016; Smith and Lewis 2011). Organisational paradox refers to tensions between collaboration and competition, routine and change, and empowerment and direction (Smith and Lewis 2011). These tensions arise because organisations are complex systems characterised by competing designs, structures, practices and processes to achieve desired outcomes. This paradox occurs at the macro level and usually during periods of organisational change and reorganisation (Smith and Lewis 2011; Ozanne et al. 2016). The learning paradox occurs when systems change, innovate and renew (Smith and Lewis 2011). This paradox addresses the tensions faced by organisations undergoing an innovation process and the resulting trade-offs between existing and new activities (Anzivino et al. 2021; Smith and Lewis 2011). These tensions also address the need for organisations to keep pace with contextual change (Ozanne et al. 2016).

Complexity science (Holland 2014) and paradox theory recognise the complexity of paradoxical tensions, which cannot be resolved by simple, linear solutions or by choosing one side of the contradiction over the other. Instead, they require a more nuanced and systemic approach (Smith et al. 2017). In this context, resilience is not an option (Walker et al. 2004), the only way to remain viable is through metamorphosis (De Toni et al. 2023), a phase transition of the service ecosystem (Polese et al. 2021).

Managing Paradoxical Tensions

Management research has paid increasing attention to this concept of tensions, which represent relationships between two different paradoxical poles (Wannags and Gold 2020).

In managing tensions, managers could respond in several ways: they could choose one of the opposing poles, but in this either/or approach, the persistent characteristics of a tension could only support a short-term solution and the contradiction could reoccur, or there is no choice that could resolve the tension (Vafeas and Hughes 2020). Managers could also compromise by viewing the opposite poles as extremes of a continuum, but this approach might not stimulate organisational vibrancy (Vafeas and Hughes 2020). Another approach deals with contingency, if/then, and needs to recognise when one condition is more appropriate than another. In a paradoxical perspective, as noted above, both poles are engaged simultaneously and, as Lewis (2000) notes, long-term sustainability involves new ways of managing multiple and divergent demands (Vafeas and Hughes 2020) from heterogeneous actors.

Furthermore, many scholars focused on tensions (Hahn et al. 2015, 2018; Haffar and Searcy 2017; Ozanne et al. 2016): similar to organisational theory, tensions were initially managed with a consistency approach (Wannags and Gold 2020). Using a paradoxical lens to investigate is useful for scholars to consider the complexity of complex problems, such as those related to sustaianability, the intrinsic value of social and environmental elements, and their systemic nature. When companies deal with economic, social, and environmental sustainability, given the complexity and dynamism of the context, the acknowledgment of inherent tensions is crucial. It allows organizational actors to understand and accept that such tensions are not only inevitable but also can and should coexist (Ozanne et al. 2016). This recognition, in turn, enables the pursuit of potential strategies for managing these tensions effectively (Hahn et al. 2015; Smith and Lewis 2011). The advantage of adopting a paradox lens is that it highlights opportunities to identify and manage the inherent paradoxes faced by firms striving to achieve complex objectives. This approach provides fertile ground for innovation and creative problem-solving (Ozanne et al. 2016; Hahn et al. 2015). In contrast to traditional approaches, the paradox approach posits that organizations should embrace the existence of paradoxical tensions. By doing so, they can manage these tensions in a way that does not necessitate compromising on certain objectives (Smith and Lewis 2011). Managing tensions implies the identification and understanding of tensions among economic, social and environmental aspects.

After this recognition phase managers could think of different strategies to manage tensions considering organization's objectives. These tensions are present at different levels in the organization and they need to be managed to achieve integrative outcomes (Hahn et al. 2015). Once tensions have been detected and paradoxical framed, three possible strategies emerge: acceptance, separation and synthesis (Hahn et al. 2015).

Since paradoxical tensions cannot be resolved by simple, linear solutions or by choosing one side of the contradiction over the other. Instead, they require a more nuanced and systemic approach (Smith et al. 2017). In this context, resilience is not an option (Walker et al. 2004), the only way to remain viable is through metamorphosis (De Toni et al. 2023), a phase transition of the service ecosystem (Polese et al. 2021).

A Framework for Matching Complexity, Frictions, and Evolution in Service Ecosystems

To advance our understanding of the evolution of service ecosystems characterized by complexity and frictions in markets and marketing, we propose the framework shown in Fig. 3.4. In particular, we deepen the concept of complexity between order and chaos by identifying three types of complexity and highlighting the role of frictions, emergence, and institutionalization in the process of service ecosystem evolution. In doing so, we leverage a classification and the "complexity expansion process" provided by De Toni et al. (2023), who described three stratified forms of social and productive organization continually changed by new technologies. We adopt such classification and process to service ecosystems, given that they both share similar roots in complexity theory. However, we update such a process to consider the actor-dependency of complexity, integrating into it the mechanisms of emergence and institutionalization and highlighting the role of frictions and their possible management. We illustrate the three complexity states using examples from healthcare systems intertwined with digitalization and new technologies.

Types of Complexity States of Service Ecosystems and Related Frictions with Illustrations

The Ordered Complexity

The first type of complexity, close to the state of order, is the "*ordered complexity*". Here, as in the case of "complication" described by Barile and Saviano (2018), the actor perceives the context of the service ecosystem (or a part of it), with its emergent properties and the possible frictions as something that can be understood, organized, embedded in known interpretation schemes, rules, and norms to co-create value.

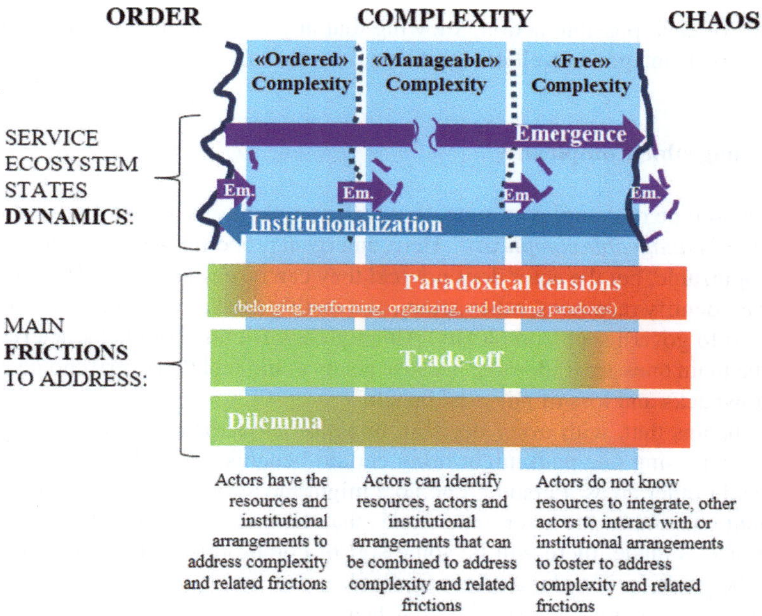

Fig. 3.4 The complexity states of service ecosystems, the related frictions and the possible service ecosystem dynamics towards evolution elaboration. *Source* The authors'

The main friction detectable here is the dilemma, since each actor, based on the perception of benefits and other preferences can make his/her own choice to better exploit the possibilities offered by the context.

Illustration: **Diabetes glucose monitoring**. In the realm of diabetes, new continuous glucose monitoring devices significantly enhance diabetes management, providing substantial benefits to both patients and the healthcare system. Indeed, measuring blood glucose levels can allow for personalized real time interventions. Wearable devices for self-monitoring allow patients to maintain a constant dialogue with their bodies, sharing and storing data on physical parameters via their smartphones. However, such technologies require skin sensors that might be uncomfortable. Moreover, such sensors might be visible to other people than the patient, preventing his/her privacy and could make certain activities difficult for patients. Deciding whether to use a device requires a choice whereby one gains some benefits, such as constant monitoring, but loses others, e.g. the inability to perform certain activities, privacy related to the chronicity of the disease, etc. The wide range of devices developed in this field, on the other side, ask doctors to be able to select the most appropriate technology for each diabetic patients.

This is an example of ***dilemma*** because patients have to choose to adopt this innovative solution or not (the Shakespearean "to be or not to be") and this decision creates a challenging scenario for them because it forces the actor to weigh the pros and cons of each possible action, knowing that any decision made will come with significant advantages as well as significant drawbacks.

The Manageable Complexity

The second type of complexity state, in which complexity is higher than in the first one, is the "*manageable complexity*". Here, emergent properties and possible frictions are recognizable, but the actor does not feel they can be easily solved. However, the actor can identify resources, other actors, and institutional arrangements that can be combined to govern such complexity. Although any friction could characterize this state, the main ones are trade-offs, because actors could feel that every choice comes with drawbacks and loss of potential benefit.

This means that, with every decision or action an actor takes, there are consequences: choosing one path might bring certain benefits but at the same time lead to losses in other areas. Pursuing one goal might mean sacrificing another, equally important one and so the actor understands that there is no perfect solution—whatever choice is made, there will be some sort of compromises, and something will have to be given up in order to move forward.

In the fast-moving context of digital health, patients often perceive that technologies prioritize market trends and the latest fashions over their actual needs and personal experiences. This can lead to solutions that do not fully address the practical challenges faced by those who use them daily. To cope with this, developers and startuppers should, first of all, try to think like a patient, analysing the problem from the patient's perspective.

Illustration: **Wearable devices for healthcare self-monitoring**. An interesting case is the high-tech T-shirts designed for self-monitoring at home by an Italian Hospital. This T-shirt is embedded with special sensors that allow key health parameters to be monitored. Upon discharge, patients are provided with these T-shirts, preconfigured to allow seamless data reading and transmission with a portable router. Then, patients wear the T-shirt comfortably at home, and it continuously collects and transmits data. Additionally, each patient receives a portable router, enabling data transmission without relying on their phone connection or home network. The T-shirt is not a limiting device, but it is much more discreet than, for example, the Holter. It is much more comfortable but still represents a constraint for patients who have to take it and bring it back to the hospital and it has to be worn for a certain period of time. This technology enables remote monitoring of patients, reducing the need for hospital visits unless concerning parameters are detected. It provides greater security for patients, who feel reassured by advanced technology during the physically and psychologically delicate post-discharge phase. It also promotes better self-monitoring, encouraging patients to gradually resume their daily activities within the comfort of their homes. Moreover, it allows for the slightly earlier discharge of stable patients, enhancing their sense of safety and independence.

In other words, although healthcare service ecosystems pervaded by new technologies may appear easy to manage, as in an "ordered complexity state", most of the actors that daily experience them will confirm that there are multiple drawbacks and dark sides of technologies, especially when related to not digitally advanced patients and doctors, or fragile patients, or technologies that have not demonstrated yet their reliability and trustworthiness. Thus, emergent properties may easily manifest, and frictions, mainly in terms of **trade-offs**, will appear because, among others, patients sustain a sacrifice knowing that they are obtaining a benefit. Patients may experience a sense of compromise or sacrifice; they are aware that using new technologies could bring them certain benefits, but they must also endure certain difficulties or inconveniences. Whether it's struggling with unfamiliar technology, navigating systems that may not be user-friendly, or facing reliability issues, patients often trade off their comfort or security for potential health improvements. This leads to a situation where technological benefits are recognized, but not without significant challenges and compromises.

The Free Complexity

The third type of complexity state, in which the complexity is the highest, is the "*free complexity*", and it is bounded by the edge of chaos. This is the area emergent properties and possible frictions may be not recognizable, nor the resources to integrate, other actors to interact with, and the institutional arrangements to rely on, promote or trying to establish are unknown yet. This is the realm of multiple frictions, and most of them may appear as paradoxical tensions.

Illustation: **Human–machine interactions in healthcare**. In the relationship between patient and doctor, and also among doctors, digital providers, and engineers, all of whom are integrating their resources within the digital healthcare service ecosystem, ***paradoxical tensions*** may also emerge. These tensions arise as each actor brings their own unique set of skills, perspectives, and expectations: patients may seek personal, empathetic care from their doctors, while doctors strive to provide this care efficiently with the aid of advanced digital tools (paradox of belonging). Meanwhile, digital providers and engineers focus on developing and implementing these technological solutions, aiming to enhance healthcare delivery and the well-being of society as a whole. However, these diverse goals and approaches can sometimes lead to conflicting priorities: patients might feel overwhelmed or disconnected by the increasing reliance on technology, preferring more direct human interaction (performing paradox). Doctors might struggle to balance the demands of maintaining a personal touch with the necessity of integrating new digital tools into their practice. Digital providers and engineers, on the other hand, may face challenges in understanding the nuanced needs of patients and healthcare providers, as their primary focus is on the technical aspects of innovation. Heterogeneous actors must navigate these tensions, trying to find ways to collaborate effectively while addressing the unique challenges posed by their differing perspectives and objectives and to foster new ways of thinking and collaborating that lead to the creation of both physical and virtual platforms where diverse actors can come together to better address patients' needs (learning paradox).

Service Ecosystems States Dynamics

The service ecosystem states do not last forever. First of all, an actor interested in performing institutional work to create, maintain, and change institutional arrangements (Lawrence and Suddaby 2009) is able to leverage its knowledge endowment (Gaeta et al. 2019) to exploit the ordered complexity, enacting the new properties to co-create value and establishing recurrent patterns of resource integration and service for service exchange. Such an act of ordering the complexity is represented by the "institutionalization" row which goes from the chaos to the order state and, in this case, from the ordered complexity to order. Similarly, in the manageable and in the free complexity, institutional work could be put in place to move from this state to the ordered one and transform a service ecosystem in a state of free to a state of manageable or ordered complexity. Here, there will be greatest exploration of new possibilities.

Secondly, interacting actors (purposefully or not) may give rise to new emergent properties (Peters 2016; Taillard et al. 2016), and complexity could expand due to waves of emergent properties from the ordered state to the chaos one (see the row of emergence that goes from order to chaos, and the resulting dotted lines of Fig. 3.4), with the edge of chaos continuously moving further into the chaos state. Specifically, all the orders of emergence shown in Vargo et al. (2023) can manifest both in the ordered and in the complex state of the service ecosystem, and the emergence and institutionalization are seen as drivers for the service ecosystem evolution through the states.

Finally, actors' perception could escalate from one actor to a community and become institutionalized at higher levels of the service ecosystem, or it could be kept secret from the other actors (at least for a while) to take advantage of the changes (Sarno et al. 2025a) as a first mover, before changing again.

Conclusions

This manuscript contributes to show that both complexity and frictions are characterized by an inner duality. Although they might seem inhibitors of value co-creation and a source of dissolution of the service ecosystems, they can also serve as catalysts for novelty, driving actors to develop new solutions and adapt to changing conditions, towards a service ecosystems evolution.

In particular, the manuscript focuses on such intricate dynamics, which can be crucial for innovation, resilience, and sustainability (among others). It offers a framework for understanding the phenomena and a typology of service ecosystem complexity states (ordered, manageable, and free complexity) based on actors' perceptions, and connects them to the most common frictions (dilemmas, trade-offs, and paradoxical tensions) found in the literature, and the main systems dynamics (emergence and institutionalization, due to actors' interactions).

The proposed typology can assist decision-makers and scholars in labeling contexts, both for sharing an intersubjective view among collaborators (Sarno et al. 2025b), and for tracing the perception of other actors depending on their knowledge endowment (Gaeta et al. 2019). Indeed, the service ecosystem state perception can likely bring to exploitation of current conditions (ordered complexity), exploration of opportunities (manageable complexity), or unveiling of new situations (free complexity). The study connects the complexity states to the main frictions identified in the literature that could be experienced by actors. Such association allows to leverage managerial studies related to selection of strategies to cope with frictions and the dependent complexity.

A specific focus on tensions, for example, can reveal a complex interplay between paradoxical poles that actors face. The traditional approaches to managing tensions—such as choosing one pole, compromising, or applying a contingency-based strategy—often provide only temporary solutions or fail to stimulate long-term organizational vibrancy (Vafeas and Hughes 2020). Instead, adopting a paradoxical lens enables managers to embrace and navigate these tensions more effectively. This approach recognizes that both poles of a paradox can coexist and that managing these tensions requires innovative and nuanced strategies (Lewis 2000; Smith and Lewis 2011).

Furthermore, acknowledging inherent tensions, particularly in complex domains (Ozanne et al. 2016; Hahn et al. 2015), and embracing paradoxical tensions can allow actors to identify opportunities for innovation and creative problem-solving, rather than merely opting for compromise or consistency (Hahn et al. 2018; Smith and Lewis 2011). Managers can employ strategies such as acceptance, separation, and synthesis to address these tensions in a manner that aligns with organizational objectives and promotes integrative outcomes (Hahn et al. 2015).

Ultimately, the ability to manage paradoxical tensions effectively is crucial for organizational resilience and viability. As complexity science suggests, service ecosystems must undergo metamorphosis—a transformation that embraces tensions (De Toni et al. 2023; Polese et al. 2021). This systemic approach to managing paradoxical tensions not only addresses immediate challenges but also fosters long-term sustainability and dynamism within service ecosystems (Walker et al. 2004).

Focusing on managerial implications, managers should recognize and manage the different types of complexity. Specifically, in the "ordered complexity" state managers should focus on the optimization of known processes and try to address frictions—mainly dilemmas—trough informed decision making. In the "manageable complexity"—where trade-offs are more evident—managers need to balance goals and resources to minimize drawbacks while maximizing benefits. In the "free complexity"—mainly addressing paradoxical tensions—managers must adopt flexible and adaptive strategies to navigate evolving challenges and harness new opportunities.

Seeing complexity and frictions as catalysis for novelty can have practical implications too. Coming back to our illustrations, for example, dilemmas in the adoption of patient-centered technologies can serve as a catalyst for the development of more intuitive and accessible solutions tailored to improve the user experience. Similarly, the trade-offs inherent in implementing digital health technologies can lead to more nuanced, patient-centered designs by exposing the gaps between technological capabilities and patient needs. Therefore, it is essential for managers to cultivate an organizational culture that recognizes and embraces these frictions as catalysts for creative problem solving and continuous system refinement.

Finally, looking at the attempts of managers to orient the evolution of service ecosystems, practitioners should pay particular attention to the dynamics among service ecosystems states presented in our framework. They could try to foster emergence (Vargo et al. 2023) to move from more predictable but already exploited states of ordered complexity to new forms of free complexity. On the other side, viability will be also pursued through institutional work to institutionalize best practices and establish robust routines and norms where manageable or free complexity was dominating. These processes support not only the smooth integration of new technologies and processes, but also the actors' ability to adapt and evolve with emerging changes in the environment. Managers play a key role in this adaptive process by fostering resilience and readiness for continuous change. They must also remain vigilant to ongoing developments in the ecosystem, which can be effectively monitored through the establishment of collaborative platforms that can leverage heterogeneous actors' dynamic capabilities and learning to navigate the complexities and frictions of the ecosystem.

Future works, among others, may be focused on deepening the role of paradoxical tensions in the sustainability transitions of service ecosystems, and the capabilities required to detect and face complexity and frictions.

References

Akaka, M.A., Chandler, J.D.: Reframing exchange: a service-ecosystems perspective. The Sage Handbook of service-dominant logic, pp. 135–148 (2019)

Anzivino, A., Galli, M., Sebastiani, R.: Addressing tensions and paradoxes in sustainable wine industry: the case of the association "Le Donne del Vino." Sustainability **13**(8), 4157 (2021)

Arthur, W.B.: Increasing returns and path dependence in the economy. University of Michigan Press (1994)

Badinelli, R., Barile, S., Ng, I., Polese, F., Saviano, M., Di Nauta, P.: Viable service systems and decision making in service management. J. Serv. Manag. **23**(4), 498–526 (2012)

Banoun, A., Dufour, L., Andiappan, M.: Evolution of a service ecosystem: longitudinal evidence from multiple shared services centers based on the economies of worth framework. J. Bus. Res. **69**(8), 2990–2998 (2016)

Barile, S., Saviano, M.: Complexity and sustainability in management: insights from a systems perspective. In: Barile, S., Pellicano, M., Polese, F. (eds.) Social Dynamics in a Systems Perspective, pp. 39–64. Springer, Cham, DE (2018)

Barile, S., Lusch, R., Reynoso, J., Saviano, M., Spohrer, J.: Systems, networks, and ecosystems in service research. J. Serv. Manag. **27**(4), 652–674 (2016)

Barile, S., Polese, F., Pels, J., Sarno, D.: Complexity and governance. Global encyclopedia of public administration, public policy, and governance. Springer (2018)

Barile, S.: Systems Thinking and Viable Systems, 1st edn, Routledge. (2024). https://doi.org/10.4324/9781003475187

Beer, S.: The viable system model. Viable Systems Model, Wiley, Chicester, (1989)

Boltanski, L., Thévenot, L.: On justification: economies of worth. Princeton University Press (2006)

Brozovic, D., Tregua, M.: The evolution of service systems to service ecosystems: a literature review. J. Manag. Rev. **24**(4), 459–479 (2021)

Cherrier, H.: Barriers to resolving institutional complexity in access-based service innovation. J. Retail. Consum. Serv. **81**, 104010 (2024)

Cheung, L., McColl-Kennedy, J.R.: Resource integration in liminal periods: transitioning to transformative service. J. Serv. Mark. **29**(6/7), 485–497 (2015)

De Toni, A.F., Zollo, G., De Zan, A.: Organizational paradoxes and metamorphosis in collective action. Systems **11**(5), 241 (2023)

Desai, D.A.: Co-creating learning: insights from complexity theory. Learn. Organ. **17**(5), 388–403 (2010)

Edvardsson, B., Kleinaltenkamp, M., Tronvoll, B., McHugh, P., Windahl, C.: Institutional logics matter when coordinating resource integration. Mark. Theory **14**(3), 291–309 (2014)

Gaeta, M., Loia, F., Sarno, D., Carrubbo, L.: Online social network viability: misinformation management based on service and systems theories. Int. J. Bus. Manag. **1**(1), 17–35 (2019)

Gallan, A.S., McColl-Kennedy, J.R., Barakshina, T., Figueiredo, B., Jefferies, J.G., Gollnhofer, J., Winklhofer, H.: Transforming community well-being through patients' lived experiences. J. Bus. Res. **100**, 376–391 (2019)

Golinelli, G.M.: Viable systems approach (VSA). Governing business dynamics (2010)

Greenwood, R., Raynard, M., Kodeih, F., Micelotta, E.R., Lounsbury, M.: Institutional complexity and organizational responses. Acad. Manag. Ann. **5**(1), 317–371 (2011)

Gummesson, E., Mele, C., Polese, F.: Complexity and viability in service ecosystems. Mark. Theory **19**(1), 3–7 (2019)

Haffar, M., Searcy, C.: Classification of trade-offs encountered in the practice of corporate sustainability. J. Bus. Ethics **140**, 495–522 (2017)

Hahn, T., Figge, F., Pinkse, J., Preuss, L.: Trade-offs in corporate sustainability: you can't have your cake and eat it. Bus. Strateg. Environ. **19**(4), 217–229 (2010)

Hahn, T., Pinkse, J., Preuss, L., Figge, F.: Tensions in corporate sustainability: towards an integrative framework. J. Bus. Ethics **127**, 297–316 (2015)

Hahn, T., Figge, F., Pinkse, J., Preuss, L.: A paradox perspective on corporate sustainability: descriptive, instrumental, and normative aspects. J. Bus. Ethics **148**, 235–248 (2018)

Harari, Y.N.: Sapiens. A brief history of humankind. Harper (2015)

Holland, J.H.: Complexity: A very short introduction. OUP Oxford (2014)

Lewis, M.W.: Exploring paradox: toward a more comprehensive guide. Acad. Manag. Rev. **25**(4), 760–776 (2000)

Lewis, M.W., Smith, W.K.: Paradox as a Metatheoretical perspective: sharpening the focus and widening the scope. J. Appl. Behav. Sci. **50**(2), 127–149 (2014)

McColl-Kennedy, J.R., Cheung, L., Coote, L.V.: Tensions and trade-offs in multi-actor service ecosystems. J. Bus. Res. **121**, 655–666 (2020)

Olsen, O.E., Kruke, B.I., Hovden, J.: Societal safety: concept, borders and dilemmas. J. Contingencies Cris. Manag. **15**(2), 69–79 (2007)

Ozanne, L.K., Phipps, M., Weaver, T., Carrington, M., Luchs, M., Catlin, J., Williams, J.: Managing the tensions at the intersection of the triple bottom line: A paradox theory approach to sustainability management. J. Public Policy Mark. **35**(2), 249–261 (2016)

Peters, L.D.: Heteropathic versus homopathic resource integration and value co-creation in service ecosystems. J. Bus. Res. **69**(8), 2999–3007 (2016)

Pole, M.S., Van de Ven, A.H.: Using paradox to build management and organization theories. Acad. Manag. Rev. **14**(4), 562–578 (1989)

Polese, F., Payne, A., Frow, P., Sarno, D., Nenonen, S.: Emergence and phase transitions in service ecosystems. J. Bus. Res. **127**, 25–34 (2021)

Putnam, L.L., Fairhurst, G.T., Banghart, S.: Contradictions, dialectics, and paradoxes in organizations: a constitutive approach. Acad. Manag. Ann. **10**(1), 65–171 (2016)
Reynoso, J., Barile, S., Saviano, M., Spohrer, J.: Service systems, networks and ecosystems: connecting the dots concisely from a systems perspective. Handbook of Service-Dominant Logic; Vargo, SL, Lusch, RF, Eds (2018)
Sarno, D., Pels, J., Polese, F., Sorhmmar, D., Tronvoll, B., Vargo, S.L.: Exploring radical market changes as phase transitions of service ecosystems: Insights from Complexity Science. In: Sarno, D., Vargo, S.L., Polese, F. (eds.). Complexity and emergence in market ecosystems. Chapter 1. Springer, New economic windows (2025a)
Sarno, D., Polese, F., Megaro, A.: Using a systems perspective for sharing market views and changes for market shaping. In: Sarno, D., Vargo, S.L., Polese, F. (eds.) Complexity and emergence in market ecosystems. Chapter 2. Springer, New economic windows (2025b)
Siltaloppi, J., Koskela-Huotari, K., Vargo, S.L.: Institutional complexity as a driver for innovation in service ecosystems. Serv. Sci. **8**(3), 333–343 (2016)
Sklyar, A., Kowalkowski, C., Sörhammar, D., Tronvoll, B.: Resource integration through digitalisation: a service ecosystem perspective. J. Mark. Manag. **35**(11–12), 974–991 (2019). https://doi.org/10.1080/0267257X.2019.1600572
Smania, G.S., Ayala, N.F., Coreynen, W., Mendes, G.H.: Data-related tensions in digital servitization ecosystems: a systematic literature review. Ind. Mark. Manage. **123**, 31–48 (2024)
Smith, W.K., Lewis, M.W.: Toward a theory of paradox: A dynamic equilibrium model of organizing. Acad. Manag. Rev. **36**(2), 381–403 (2011)
Smith, W.K., Erez, M., Jarvenpaa, S., Lewis, M.W., Tracey, P.: Adding complexity to theories of paradox, tensions, and dualities of innovation and change: introduction to organization studies special issue on paradox, tensions, and dualities of innovation and change. Organ. Stud. **38**(3–4), 303–317 (2017)
Sundaramurthy, C., Lewis, M.: Control and collaboration: paradoxes of governance. Acad. Manag. Rev. **28**(3), 397–415 (2003)
Taillard, M., Peters, L.D., Pels, J., Mele, C.: The role of shared intentions in the emergence of service ecosystems. J. Bus. Res. **69**(8), 2972–2980 (2016)
Teece, D., Pisano, G.: The dynamic capabilities of firms, pp. 195–213. Springer, Berlin Heidelberg (2003)
Thornton, P.H., Ocasio, W., Lounsbury, M.: The institutional logics perspective: a new approach to culture, structure and process. OUP Oxford (2012)
Thornton, P.H., Ocasio, W.: Institutional logics and the historical contingency of power in organizations: executive succession in the higher education publishing industry, 1958–1990. Am. J. Sociol. **105**(3), 801–843 (1999)
Tóth, Z., Sklyar, A., Kowalkowski, C., Sörhammar, D., Tronvoll, B., Wirths, O.: Tensions in digital servitization through a paradox lens. Ind. Mark. Manage. **102**, 438–450 (2022)
Vafeas, M., Hughes, T.: Resource integration: adopting a paradox perspective to inform the management of tensions in customer resource allocation. Ind. Mark. Manage. **91**, 596–609 (2020)
Vargo, S.L., Lusch, R.F.: It's all B2 and beyond: Toward a systems perspective of the market. Ind. Mark. Manage. **40**(2), 181–187 (2011)
Vargo, S.L., Lusch, R.F.: Institutions and axioms: an extension and update of service-dominant logic. J. Acad. Mark. Sci. **44**, 5–23 (2016)
Vargo, S.L., Lusch, R.F.: Service-dominant logic 2025. Int. J. Res. Mark. **34**(1), 46–67 (2017)
Vargo, S.L., Akaka, M.A., Wieland, H.: Rethinking the process of diffusion in innovation: a service-ecosystems and institutional perspective. J. Bus. Res. **116**, 526–534 (2020)
Vargo, S.L., Peters, L., Kjellberg, H., Koskela-Huotari, K., Nenonen, S., Polese, F., Sarno, D., Vaughan, C.: Emergence in marketing: an institutional and ecosystem framework. J. Acad. Mark. Sci. **51**(1), 2–22 (2023)

Vink, J., Koskela-Huotari, K., Tronvoll, B., Edvardsson, B., Wetter-Edman, K.: Service ecosystem design: propositions, process model, and future research agenda. J. Serv. Res. **24**(2), 168–186 (2021)

Walker, B., Holling, C.S., Carpenter, S.R., Kinzig, A.: Resilience, adaptability and transformability in social–ecological systems. Ecol Soc **9**(2) (2004)

Wannags, L.L., Gold, S.: Assessing tensions in corporate sustainability transition: From a review of the literature towards an actor-oriented management approach. J. Clean. Prod. **264**, 121662 (2020)

Chapter 4
Socio-Economic Ecosystems Seen as Process of Complexification

Petter Braathen

Abstract This chapter explores socio-economic ecosystems as processes of complexification. A conceptual model is introduced where isomorphic structures of complexity grow within nested hierarchies, progressing from individuals to practices and routines, organizations, and market systems. Rather than viewing these systems through a reductionist lens—as static entities or aggregates of resources—this analysis emphasizes how structures emerge through relational dynamics and interdependencies. Drawing on complex adaptive systems theory, we propose a framework based on a relational ontology of process that offers a systemic-evolutionary perspective on how socio-economic systems emerge and adapt. The study positions complexity theory as a versatile framework applicable across disciplines. Its systemic-evolutionary approach provides a unifying lens for fields such as management, organizational theory, and service theory, facilitating an integrated perspective that spans traditional disciplinary boundaries. Such transdisciplinarity resonates with a growing body of work advocating for frameworks that can address the complex, interconnected, and dynamic nature of modern socio-economic phenomena.

Keywords Complexity · Complex adaptive systems · Service ecosystems

Introduction

It is generally accepted that simpler, smaller components predate more complex, composite systems. Thus, societal evolution tends to generate increasingly complex systems, gradually adding more levels to a hierarchy. For example, individuals form families, which combine to create communities, which then integrate into cities, nations, and, ultimately, global societies. The rationale behind this sequential complexification is evident: higher-order social structures can only be constructed from simpler ones because these simpler structures provide essential building blocks. Hence, there is a directional evolutionary process whereby higher-order systems

P. Braathen (✉)
Inland Norway University, Kongsvinger, Norway
e-mail: petter.braathen@inn.no

D. Sarno et al. (eds.), *Complexity and Emergence in Market Ecosystems*, New Economic Windows, https://doi.org/10.1007/978-3-031-81942-1_4

come into being with new properties, with a fundamental increase in complexity from lower-order systems—in other words, a *complexification process.*

Investigating the mechanisms behind these evolutionary transitions and determining which levels emerged at certain times and for what reasons is an intricate process. A classical reductionist approach would address a phenomenon to analyze a mutually exclusive and collectively exhaustive description of its elements. While this would provide valuable insights into the nature of a system's elements, it would be impossible to describe the phenomenon as a whole from the properties of its constituent parts or their aggregate (Heylighen et al. 2006). On the contrary, propositional properties of an emergent phenomenon are not based on the ontological substance of elements, but rather on the processes of organizing that take place, that is, the *interactions* between elements. Therefore, to comprehend the emergence of new phenomena, it is essential to concentrate on the relational ontological aspects of such emergence, further accompanied by an evolutionary epistemology that addresses change.

Complexity may be defined by two foundational categories, *variety* and *relation* (Heylighen et al. 2006; Holland 2014). In his seminal article "The architecture of complexity", Herbert Simon (1962) explains how complexity forms natural nested hierarchies composed of subsystems, from the global ecosystem to societies to human individuals, extending down to the simplest elements of physical nature. Hence, both markets and organizations may be seen as complex systems in a nested hierarchy (Arthur, 1999). This isomorphic similarity has inspired management scholars to adopt a *systemic* (Anderson 1999; Beer 1995; Lusch and Vargo 2014) and *evolutionary* (Hannan and Freeman 1977; Nelson and Winter 1985) perspective to understand socio-economic phenomena.

Across the different systemic approaches, we find these same foundational ontological units that constitute complexity, *variety and relation.* In general systems theory (GST) interdependency among a variety of elements emerge into systems as a whole (Von Bertalanffy 1972), branching into theory of complex adaptive systems (CAS) where local interactions between a variety of actants can explain systemic behavior (Holland 2014), and evolutionary theory explains how relations are formed as a consequence of selection processes among a variety of actors (Hannan and Freeman 1977). This general ontological applicability and systemic-evolutionary epistemology open the possibility for transdisciplinary application across several different sub-domains.

Systems perspectives have attracted attention from management scholars (Anderson 1999; Cobben et al. 2022; Jacobides et al. 2018; Lusch and Vargo 2014). The conceptual model of a business ecosystem has been introduced, inspired by the category used to describe biological systems (Moore 1997), and subsequently applied to management, marketing, strategy and organization literature (Cobben et al. 2022). A business ecosystem is constituted by complementarities, variety, that are dynamically interdependent, thus related (Jacobides et al. 2018), where the constant recombination of its resources "results in a commonly created output based on the ecosystem's value proposition" (Cobben et al. 2022, p. 38). Also in service theory we find well-established perspectives from theory of complex adaptive systems (Barile and Polese

2010; Lusch and Vargo 2014). Service-dominant logic is particularly coherent with the ontology of complexity where the operant and operand resources corresponds to the variety, and where resource integration through interaction, corresponds with the dynamic relational aspect (Lusch and Vargo 2014).

Hence, we see how the core ontological units of complexity, that is variety (e.g. resources, elements, actants, processes) and relations (e.g. interaction; physical, temporal or hierarchical interdependencies), are foundational to a systemic-evolutionary perspective. Such transdisciplinary approach supports interdisciplinary frameworks such as for example, theory of service ecosystems (Vargo et al. 2023), theory of routines and capabilities (Dosi et al. 2001; Feldman et al. 2016), or theory of institutions (Scott 2013). This analysis aims to explore this relationship and extend current understanding of how complexity is a foundational driver for any socio-economic (eco)systems.

Theoretical Foundation

Complexity and the Need for Inter- and Trans-Disciplinarity

The partitioning of the disciplines is a response to practical observations of the real world (Klein 2014). The natural sciences mirror the emergent logical levels of nature, from physics, to chemistry, to biology, whereas many other disciplines address practical needs in society; e.g. law <–> lawyer (to help solve disputes) or architecture <–> architect (design buildings). While the traditional disciplines are still highly relevant, we are increasingly faced with challenges of a different nature. The world's largest problems are complex, as for example, over-population, climate changes, or sustainable food production, and cannot be described adequately by any of the traditional disciplines, although all of them make important contributions. It is not sufficient to use a multidisciplinary approach that only to become serial and additive (Klein 2014). An interdisciplinary approach, on the other hand, attempts to integrate, synthesize, and restructure categories to unify different forms of knowledge. Transdisciplinary perspectives go further provide general descriptions across different domains.

Figure 4.1 is an attempt to illustrate this situation. On one axis, we find the duality (and spectrum) between theory and practice: in other words, the degree of abstraction, from the practical observation of reality, to the forming of theoretical categories evolving into structures of knowledge and theory. On the other axis, we find the degree of category complexity, that is the variety of categories and their interrelatedness needed to describe the phenomenon. This dimension implies the duality between, on the one hand, the traditional disciplinary partitioning of knowledge, and on the other, a connected structure of interrelated categories.

The consequent four quadrants portray the challenge we are currently faced with. In the two lower quadrants, we see the traditional correspondence between the challenges of the practical world and the corresponding established academic disciplines.

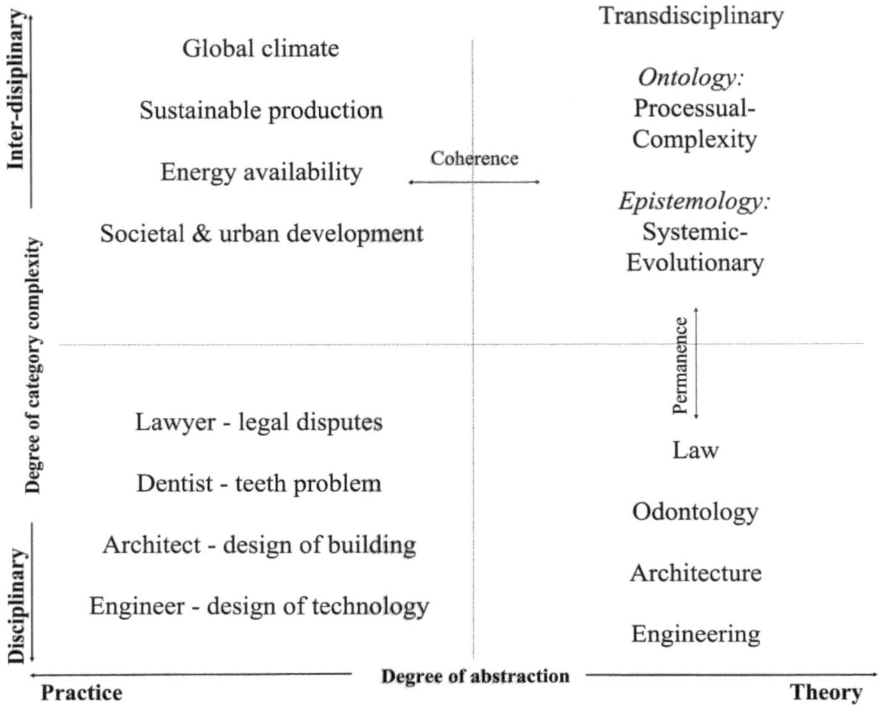

Fig. 4.1 The role of transdisciplinary systemic-evolutionary perspective. *Source* Author's elaboration

The upper left quadrant addresses current actual phenomena of a complex nature, constituted by several interrelated sub-phenomena. The upper right quadrant illustrates the need for a corresponding theoretical domain that again implies development of transdisciplinary ontology and epistemology. Further, a transdisciplinary framework must both have the ability to coherently correspond with the complex structure of elements and relations in the practical phenomenon, and to build on current categories in existing disciplines to maintain permanence of existing knowledge (Klein 2014).

One such approach will be explored in this analysis,–that is an ontology of action founded in process philosophy (Hernes 2007) complemented by a systemic-evolutionary epistemology (Heylighen 2011). The starting point will be the foundational concept of a complexification process constituted by the two dimensions: (1) variety as a set of differences that implicate distinctions and (2) relations as interactions constituted by an action and the corresponding re-action. This is observed through a systemic-evolutionary lens that supports development of interdisciplinary understanding and metatheoretical frameworks. We will apply the concepts to socio-economic organizing structures as business ecosystems, organizations and institutions to demonstrate its transdisciplinary applicability.

Towards a Systemic-Evolutionary Perspective

Prevailing reductionistic and deterministic methods, while proven massively effective for economic and societal progress, fall short of capturing the intricate dynamics of real-world socio-economic issues. Reductionism, dissecting systems into their constituent parts, fails to account for the organizing dynamic processes that form the structure of a phenomenon. Determinism, on the other hand, assumes predictability based on fixed laws and causal relationships and thus ignores the inherent uncertainty and variability present in human behavior and in communities and societies at large (Heylighen et al. 2006).

In contrast, evolutionary theory offers a dynamic perspective by introducing gradual change resulting from interactions between a system and its environment (Darwin and Huxley 2003), and a more fluid understanding of how various components within a system can adapt over time when influenced by external pressures and internal dynamics (Kauffman 1996). Further, evolutionary theory recognizes individual agency and accepts variability and spontaneity as inherent features of social systems (Axelrod 1997). However, despite these advancements, evolutionary theory falls short in accounting for the emergent properties that arise in dynamic systems, as it remains rooted in reductionism, focusing on interactions between constituent material parts rather than relational organizing processes.

A systems perspective, on the other hand, recognizes emergent phenomena as a result of collective dynamics. The word system has its origins in the Greek word systema and means a whole compounded by parts (*System | Search Online Etymology Dictionary*, n.d.). In systems theory, a system is described as a collection of elements, connected through relationships, that as a whole has its own identity that distinguishes it from its environment (Heylighen 2011). General systems theory seeks to explicate "...models, principles, and laws that apply to generalized systems or their subclasses, irrespective of their particular kind, the nature of their component elements, and the relationships or 'forces' between them" (Von Bertalanffy 1972, p. 32). Systems can be of any nature, for example, physical, chemical, biological, social, or cultural, and the theory's independence from corresponding scientific disciplines enables the development of unified interdisciplinary knowledge. As such, general systems are abstract phenomena, independent of the concrete and independent of any domain ontology and corresponding epistemology. Therefore, heterogenous systems may show similarities in their organizing structures, that is, they are *isomorphic* (Heylighen 2011). For example, the material structure of artificial intelligence is different from that of the human brain, but may show isomorphic propositional properties when observed as-a-whole or when we speak of the concepts of neurons and networks. Hence, we may also apply other meta-system perspectives, such as mathematics or philosophy, across systems of different kinds, yielding possibilities for inter- and trans-disciplinary analysis (Klein 2014).

From the definition of a system above, we can see that the two ontological categories of any system are the "collection of elements" that are "connected through relationships," that is, *elements* and *relations*. Typical *elements* in a socio-economic system could be individuals, teams, communities, as well as any type of non-human resource involved in processes of value creation, as for example a computer system. The connecting *relations* of a socio-economic system would be of a processual nature, such as communication, coordination, and collaboration.

The two ontological dimensions are coherent, with more specific categories derived in diverse disciplines of social science. First is the ontology of a system's elements. In service theory, and in particular service-dominant logic (Lusch and Vargo 2014) we find operand and operant resources and actors with agency. The actors could be individuals or systems of actors and resources, depending on the level of observation. Similarly, in sociology and organization theory, we find categories of human actors and non-human actants that form organizing structures as habits, routines, capabilities, organizations (e.g., Feldman et al. 2016; Nelson and Winter 1985; Pentland et al. 2010). In strategy, we speak of resources and actors that form unique structures, which can constitute a competitive advantage (Barney 1991). In sociology, political science, organization theory, service theory, and economics, we find institutions as emergent arrangements from interacting actors, providing a logic that shapes the environmental conditions for its organizing structure (Scott 2013).

The isomorphic properties of these categories across disciplines are strikingly similar. To advance an interdisciplinary perspective, we will define the ontological elements needed to describe any socio-economic system as the set of <actors, resources, institutions>. Even though institutions are themselves systems of organized activity, they may function as an element of a system on an aggregated analytical level. The other ontological dimension of a system addresses the *relations* between elements. We depart from a process perspective found both in organization theory where routines and capabilities emerge as patterns from actants that interact (Feldman et al. 2016), and in service literature where actors co-create value by service exchange and integration of resources through interaction (Lusch and Vargo 2014).

To further advance an interdisciplinary perspective, we are informed by a 'strong process view' found in organization theory (Nayak and Chia 2011), as well as in sociology (Luhmann 1995), and founded on process philosophy (Rescher 2000). The unit of analysis is the process of organizing rather on the organizational entity (Weick et al. 2005). This performative practice perspective is primarily based on a relational ontology of the interactions that enable a repetitive pattern as a structure to come into being (Hernes 2007), or as an ongoing process of communication among actors in social systems (Luhmann 1995). Such process perspective corresponds with service-dominant logic, which moves away from goods as substance to focus on service as action and the transformative process that may arise from interactions through service exchange and co-creation among actors (Lusch and Vargo 2014). *Interactions* represent the reciprocal nature of actions by actors and resources and become the foundational category a relational ontology found in both systemic and evolutionary epistemology. *Co-creation*, on the other hand, accentuates the performative mutual interrelatedness between actors that may take the form of coordination, collaboration, and cooperation. Hence, we arrive at a set of interdisciplinary ontological categories that belong to the systemic dimension of *relations*, that is, <interaction and co-creation>.

Since we aim to study socio-economic phenomena comprised of multiple types of organizing structures, we must be able to describe hierarchies of different types, such as routines, capabilities, organizations, institutions and markets, or networks, communities, and cities in a society (Gulati et al. 2012). The business ecosystem is a common category to accommodate the need for generalized descriptions that may be applied across levels and domains. Moore, who introduced the term to the business literature, defines the business ecosystem as a "... *community supported by a foundation of interacting organizations and individuals, the organisms of the business world. This economic community produces goods and services of value to customers, who are themselves members of the ecosystem... Over time, they coevolve their capabilities and roles, and tend to align themselves with the direction set by one or more central companies*" (Moore 1997, p. 26).

Building on this, various disciplines have embraced the concept and refined the definition to address their focal interests while seeking to maintain its general applicability to different systems. In the literature on strategy and innovation, an ecosystem is defined as "*the alignment structure of the multilateral set of partners that need to interact in order for a focal value proposition to materialize*" (Adner 2017, p. 41). Scholars in organization theory have defined ecosystems as "*collective entities of organizations that are dependent on each other for value creation*" (Gupta et al. 2023). Finally, service theory defines service ecosystems as "*a system of social and economic actors that cocreate value by exchanging service*" (Tuominen et al. 2020), where the ecosystem is viewed as a self-organizing emergent entity that evolves through interactions between actors with different resources and intentions (Vargo et al. 2023).

Hence, theory of ecosystems serves as a meta-theoretical framework across different socio-economic domains with an interdisciplinary potential to integrate concepts, and a transdisciplinary potential to explain phenomena across co-existing hierarchical levels of reality. The concept of an ecosystem provides a conceptual lens that is coherent with a systemic and evolutionary epistemology of socio-economic systems, and founded on an ontology of action from a variety of actors and resources that are related through interactions. Variety and relations are also the two founding concepts in complexity, and the evolutionary process constitute a complexification process.

Socio-Economic Systems as an Evolution of Complexity

The word "complex" stems from the Latin word complector, meaning to be entwined or weaved together (*Complex | Search Online Etymology Dictionary*, n.d.). *Complex systems* belong to a subset of general systems, where complexity refers to the variety of elements in a system and how they are interrelated. The degree of complexity in a system is decided by the number and complementarity of elements that can be distinguished and the number and strength of relations that exist between them. Complexity increases when the number of distinct elements in a system increases or when the number of relations between them increases. Similarly, complexity decreases if elements are distinct, but not connected, or if they are homogenous and hence not distinctly different (Heylighen 1999). Complex systems theory is concerned with the collective behavior constituted by the set of local interactions between elements (Holland 2014), and bridges perspectives in evolutionary theory and general systems theory.

Distinctions as the first dimension of complexity are observed difference, and in the words of Gregory Bateson, the difference that makes a difference (Bateson 1972). They are observed as propositional properties of resources, agents or institutions, or their actions. Distinctions correspond to the concept of variety in evolutionary theory, and whereas variety originally referred to the diversity of traits within a population, it can also refer to variations in technology, new behavior, or ideas in a socio-economic ecosystem. Human individuals are idiosyncratic; hence, they form a variety of heterogeneous distinguishable agents. Non-human resources are similarly distinguishable by their propositional properties and the services they offer. On a scale at which we observe a global market, heterogeneity of individuals and resources may be harder to identify, and commonalities between them are more visible. This is a matter of level of observational scale and analysis, just as two different lakes look alike from an airplane, and we are unable to distinguish between the two worlds of life within them.

Relations constitute the other dimension of complexity, and from our understanding of socio-economic ecosystems they take the form of interaction or co-creation. When variety increases to a number n, the number of possible relations increases exponentially, following Metcalf's law, which states that the number of relations between the elements in a system, are $(n*(n-1))$ (Metcalfe 2013). The complexity of a system also increases when the strength of relations increases. Strength can be determined by the amount of information that is associated with an interaction, the degree of interrelatedness in an exchange, or the level of relational trust present in a co-creation process.

In an evolutionary framework, we refer to ability and willingness to relate as selection. In a selection process, a system tries out all possible relations (blind variation) between its elements. However, in socio-economic ecosystems, a system is conditioned by external influential and controlling factors such as institutions, market dynamics, and more. This guides the selection process, and only certain relations are retained, namely ones that are adequate, successful, and relevant to the context in which they occur. Therefore, we may assume that only these stable patterns of interactions are retained and selected, and according to Donald Campbell (1974), this phenomenon may be called selective retention as opposed to blind variation where every possible relation is tried out (Campbell 1974).

From this, we see that any socio-economic ecosystem may be described by the two dimensions of complexity, that is, the variety of its elements and the relations formed by retained interactions between the elements. This is coherent with the two dimensions of evolution, that is, variety and selection, and indicates that any socio-economic system evolves through a process of complexification.

Emergence from a Process of Complexification

While there is no generally accepted definition of when a system is complex, there are archetypical qualifying properties. A complex system typically shows non-linear behavior, where it may respond in different ways to the same input, depending on its state (Holland 2006). Further, a complex system may show new qualitative properties as a whole that are not apparent from the aggregate of its components. This is in line with Aristotle's expression that the whole is *more* than the sum of its parts. What constitutes "more" in this expression is referred to as emergent properties, which are qualitatively different properties that cannot be reduced to properties of the parts from which they emerge. A car has a property that it can drive, unlike any of its parts. Music shows the property of melody, unlike properties of the individual notes.

Structural nested hierarchies are formed as a result of increased complexity (Simon 1962). This process is a natural emergent process in complex systems and should not be exchanged for the designed hierarchies of power in a traditional organization. "By a hierarchic system…I mean a system that is composed of interrelated sub-systems, each of the latter being, in turn, hierarchic in structure until we reach some lowest level of elementary subsystem….it is somewhat arbitrary as to where we leave off the partitioning, and what subsystems we take as elementary" (Simon 1962, p. 468).

Emergence as a phenomenon is still a bit of a wonder and has attracted the attention of social science for more than a century. Early scholars of sociology, economics, and political science studied how routines and institutions would emerge from systems of social and economic activity (Scott 2013). Later, much focus in organization theory has been on the emergence of organizational structures, such as organizational routines (Feldman et al. 2016; Pentland et al. 2010), organizational capabilities (Helfat and Winter 2011; Nelson and Winter 1985; Parmigiani and Howard-Grenville 2011), and dynamic capabilities as meta-capabilities that change organizational capabilities (Teece et al. 1997). In service theory, scholars have studied the emergence of service ecosystems (Taillard et al. 2016), as well as emergence of institutional arrangements (Vargo et al. 2023).

New orders may emerge from both static and dynamic structures of complexity with differing degrees of governance of coherence (control) between different levels. In nature, we commonly observe the spontaneous emergence of static structures. For example, snow crystals in ice flowers on a window, where the ice flower emerges and appears as a piece of art without a designer who determines its aesthetic properties. This phenomenon arises from the self-organization of an overall order in which the ice flower evolves as a variety of heterogeneous snow crystals interact (relate) to form a unique pattern. Without the connection between the snow crystals, we would only observe random disordered snow crystals on the window. This disorder corresponds to the natural tendency of physical systems that do not exchange energy with the outside world, denoted as *entropy*. We experience the phenomenon in our daily lives, for example, as the decay and accumulation of dust in an unattended house, or the wear and tear of machines that do not self-repair.

We find similar static structures in socio-economic ecosystems. Whereas an ice flower evolves without a goal or central design, formal organizations are examples of supersystems with a "designed" static structure, which has a purpose and is motivated by hierarchical control (Jacobides et al. 2018). We may observe this static organizational structure at a certain analytical level while recognizing that we find highly dynamic systems at both levels below, for example, inside a team in an organization, and levels above, for instance, in an open market. Take, for example, a national post-service, which has evolved over hundreds of years into a complex structure of interconnected variety of elements such as post offices, distribution centers, transportation networks, and delivery personnel. Each organizational unit has a clearly defined role and responsibility, and there are standardized processes for consistent service across the system. From this static structure, properties may still emerge at the level above. These could be qualities such as reliability, social cohesion, economic connectivity, and perceived national coverage.

On the other hand, new emergent properties also arise from highly dynamic systems of complexity. All living and social systems are of such nature. These systems are not static but highly dynamic. Their elements do not necessarily follow standard processes or adhere to defined roles and responsibilities. On the contrary, uncertainty and new variety is imported into such systems from the outside world through agents, resources, and institutions that are part of other socio-economic ecosystems such as family, community, and society at large.

Hence, any socio-economic ecosystem imports "energy" with a low degree of entropy from its members. Through interactive processes within the system, this energy is transformed, releasing entropy, which then needs to be dissipated. Living systems such as plants absorb solar energy and convert it through photosynthesis, producing both chemical energy to sustain the plant and heat as a form of dissipated energy. Similarly, social-economic ecosystems, such as organizations or communities, absorb information, ideas, and motivational energy from their members, transforming these inputs through processes like interaction and co-creation, leading to emergent properties such as innovative solutions, cultural norms, or market trends. The transformation process releases entropy, such as waste, heat, energy from relational friction or synergies.

These highly dynamic systems are characterized as "dissipative systems that are far from equilibrium" (Prigogine and Stengers 1984). It is their dissipative nature (on multiple levels), where energy flows into a system and is transformed, that enables a pattern of interactions that gradually coalesce over time, forming the fabric of a new phenomenon. The establishing pattern appears with a type of dynamic stability that is crucial for our observation of emergent holistic properties.

Any propositional property of a system-as-a-whole must show some type of stability, long enough for it to be observable; otherwise, we could not distinguish its existence as a difference, a new variety, relative to the general environment. Even if the system is in constant flux and movement, it has observational emergent properties as a whole (e.g. a swirl of water that may be observed as a phenomenon), which are a result of constant interactional change. Ilya Prigogine was awarded the Nobel Prize in 1977 for his work on the dynamic stability of dissipative systems, and his theories are foundational in understanding emergent phenomena dynamic stability through constant flux in any social-economic ecosystem (Prigogine and Stengers 1984).

To summarize, the processes of complexification and evolution drive the organizing structuration of a nested hierarchy. New variety is introduced from the environment or by transformations from internal interactions, and new relations are made possible due to environmental changes or internal selection. As for any natural hierarchy, the lower levels must form before the higher levels at which emergent properties are created by the complexity of the lower levels.

Adaption and (Co)-evolution Through a Complexification Process

So far, we have described how socio-economic ecosystems form as structures of complexity and how complex dissipative structures far from equilibrium emerge with a dynamic stability based on constant change. Once a system has come into being as a distinct phenomenon that can be observed with emerging propositional properties, the question becomes how it can remain viable over time. To do so, it must both be able to remain viable to its environment, that is, through its co-creation of value with other actors, and have the ability to uphold its own organizing structure; be resilient to external and internal perturbations. In evolutionary terms, we speak about *system-environment fitness.*

In the English language, the adjective "fit" may be understood in two ways. First, as being in a state, shape, and condition, able to do something. Second, as being adapted to an environment so as to uphold existence. The first dimension refers to the internal state and fitness of a system. We can call this *structural fitness* (Heylighen, 1999). This comes from the stability of a system's internal structure and further refers to the ability of the system to reliably (re)produce its output. The second interpretation refers to the external and *functional fitness* of a system (Heylighen 1999). This is the relative fitness of a system to its environment and is determined by the system's ability to survive external perturbations and challenges, as well as by how well it is able to relate to agents, resources, and institutions present in its environment, thus upholding the system it is a part of. The two perspectives should be seen as complementary, and what is internal for one system may be external to the (sub)systems in a nested hierarchy of systems. We can describe the total fitness of any socio-economic ecosystem by the combined concepts of *structural and functional fitness.*

Ashby's *law of requisite variety* addresses the fitness of systems as a *function of complexity* (Ashby 1958). It states that for a system to effectively control or respond to a variety of states in its environment, it must have at least an equivalent amount of variety within its response (control) mechanisms. In other words, the complexity of the focal system must match the complexity of the environmental system in which it operates. This implies that unless an organization can observe and understand its environment, that is, to distinguish between the variety of agents, resources, and institutions involved and the way they interact and co-create value, it will not be able to adequately respond to challenges and opportunities or remain with a *function* in the surrounding system it is part of. In other words, according to Ashby's law, the system is left with no choice other than to adapt through an evolutionary process of complexification in order to remain functionally fit.

Complexity grows with each new element available in a system (resource, agent, institution), and with each new retained relation (interaction, co-creation). In practical terms, an organization may increase its variety by, for example, diversifying its product offerings, investing in new technologies, acquiring and integrating new competence, or adopting more agile operational practices. With such increased complexity comes potential increased repertoire and possible value propositions. An increased repertoire has the potential to increase functional fitness, given that the service is perceived as valuable by others in its environment.

In general, evolution tends to drive toward increased complexity (Heylighen 1999). This increase and the drive for both structural and functional fitness between multiple systems initiate co-evolutionary processes (Hannan and Freeman 1977). Co-evolution refers to the reciprocal evolutionary changes that occur between interacting systems driven by the need to adapt and remain fit relative to each other's adaptations and the broader environment. In this complexification process, new variety and possible relations are introduced, and the relations that are already present in a system may become weaker. With weaker relations, constraints on variety are relaxed, and new relations with other systems can be selected.

This is also true between levels in a nested hierarchy. In a structural hierarchy, as defined by H. Simon, the lower levels of the hierarchy have a stronger structural fitness (Simon 1962). It is only after the high fitness structure has formed that weaker fitness relations are tried out. For example, it takes less effort to break a school of fish than to separate the cells in the fish body. Similarly, for a social system, coworkers who interact over time form strong personal relationships and interpersonal bonds. A well-functional team of highly competent individuals will have stronger structural fitness than the capability of interacting routines in a global organization.

While the adaptive process of complexification is necessary for the viability of a system, it also introduces significant challenges, as the expansion of internal complexity comes with significant costs. Higher complexity means more options available to a system, and it becomes harder to decide which action to take. With increased complexity, the potential configurations a system has to undergo to find the right one for the situation in the environment increases exponentially. In the search process, the system runs the risk of being destroyed or not being selected in time. Hence, while a system needs to increase its variety to deal with challenges from the environment, there is also a contradictory drive to enforce a set of constraints (i.e., to simplify) to increase the speed of convergence to adequate configurations (Heylighen 1995). This contradictory situation represents a trade-off. This principle highlights the balance that must be struck between having a wide range of possible actions and the ability to make effective decisions.

Each possibility involves a certain set of interactions between a specific set of agents or resources. Hence, the interactions needed form a type of *constraint* on the system, where the constraint refers to the total potential number of all possible interactions between the agents or resources (Heylighen 1995). In other words, for an ecosystem to create a particular service, a specific interactional pattern is needed between a specific set of actors and resources. Hence, the actors are constrained from all other possible interactional patterns that would create other services. Hence, for the system to stay viable in the nested hierarchy, it will always be constrained by the needs of the system on the level above, and hence, constrain the total set of possible interactions. An increase in the number of relations enforces more constraints on the system, forming more order and structure. Equally, introducing a new variety relaxes the number of constraints and increases the complexity of the system.

To illustrate, we will use a commonly known example from the Apple corporation. Apple's renewed success came as a result of a partly planned and partly coincidental symbiosis and interaction between technology and design. The first iMacs were intended as a computer that did not look like a computer with the intention of expanding into new market segments. A new super-capability was created from the interaction between two existing capabilities: engineering and design. The super-capability would redefine how technology was perceived and how it was applied in people's lives, with a new repertoire of focal value propositions for telephones, music-playing devices, and computers. The new, emerging qualitative properties were different from both the engineering and design capabilities individually. The super-capability would also direct all future development in both engineering and design and in routines on lower levels. The required design profile and brand experience put a set of constraints on the possible repertoire of the design and engineering capabilities individually, and, as we zoom in, on the repertoire of organizational routines and on the interactions between resources and actors at lower levels. In short, it provided a hierarchy of constraints on the variety provided by actors and resources on all levels in the organization.

In summary, Ashby's Law of Requisite Variety emphasizes the necessity for systems to match the complexity of their environment to maintain viability and resilience. For socio-economic ecosystems, this involves continuously adapting both their structural and functional fitness to keep pace with evolving external conditions. This adaptive process is inherently a complexification process that requires balancing an increase of internal variety with enforcement of relational constraints to ensure timely and effective decision making.

We are now ready to define socio-economic ecosystems based on an interdisciplinary evolutionary-systemic perspective of complexity.

(i) A socio-economic ecosystem is a constraint on the variety of interactions and co-creation among agents and resources that maintains a shared set of focal value propositions in an institutional context.

More specifically, this leads to an isomorphic systemic description of the typologies of a nested organizational hierarchy:

(ii) An organizational routine is a socio-economic ecosystem emerging from a constraint on variety between agents and resources that maintain a service
(iii) An organizational capability is a socio-economic ecosystem emerging from a constraint on a variety of routines that maintains a strategic value proposition in an organization or a service ecosystem;
(iv) i.e., (a constraint on a variety on (a constraint on variety of interactions between agents and resources))
(v) An organization is a socio-economic ecosystem emerging from a constraint on variety of capabilities, which maintains a function in a service or market eco-system;

i.e., (a constraint on variety on (a constraint on variety on (a constraint on variety of interactions between agents and resources)))

Describing Socio-Economic Systems in Terms of Complexity

We will now apply principles from the analysis above to address and exemplify a set of socio-economic systems on different levels in the nested hierarchy of. The description is brief, with the only intention to illustrate how each type of system may be described by the same concepts of complexity, independent of level and ontological substance domain. A supporting illustration is provided in Fig. 4.2. Our perspective is a matter of level of observation and analytical scale.

We start with the organization as a complex system of a variety of capabilities (strategic resources) interdependently related (Barney 1991). When we zoom out, the organization may be a member of a business ecosystem where a variety of actors are related through collaborative exchange of complementary service to co-create value around a shared set of value propositions (Adner 2017; Jacobides et al. 2018; Vargo et al. 2023). Or even further out as an abstract element in a general market system where processes emerge and evolve from dynamics between a variety of interacting actors (Storbacka and Nenonen 2011).

When we zoom in, we can observe various levels of resource integrating systems. Dosi et al. (2001:4) describe it as follows: "routines (that) are units or 'chunks' of organized activity…they are the building blocks of capabilities…Individual skills, in turn, are among the building blocks of organizational routines." Hence, where organizations may be characterized by complex systems of organizational capabilities, capabilities are in turn complex systems of organizational routines, where a variety of routines are interrelated through their reciprocal service exchanged (Feldman et al. 2016; Tuominen et al. 2020). Finally, routines are observable and recognizable patterns of activity from a variety of actants related by interactions, performing the routine with effortful accomplishment (Pentland et al. 2010). Each level comes into

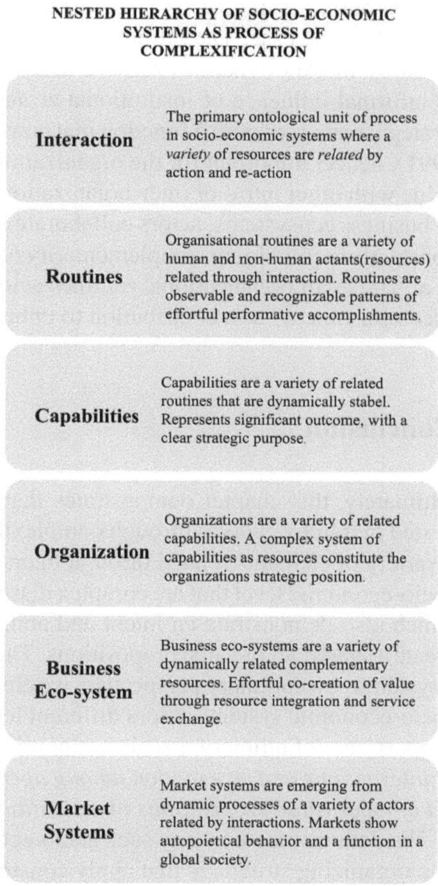

Fig. 4.2 Describing socio-economic systems in terms of complexity. *Source* Author's elaboration

being as a complexification process and emerge with new qualitative properties on the level above.

Further, at all levels we may describe the phenomena from a strong process perspective, as dynamically stable patterns of activity. Based on our analysis of the dissipative structures above, we understand that the emergence of the qualitative identity and propositional properties come from dynamic stability of on-going interactions between a variety of agents and resources. For any socio-economic system to 'exist', the pattern of ongoing activity must be observable and recognizable, distinctly different from its environment.

Finally, in order for the unit to be resilient and remain stable over time, the function of complexity must obey Ashby's law of requisite variety. This corresponds to the purpose and goal directedness of socio-economic ecosystems. Although many natural systems, such as rainforests, evolve and emerge from complexity, they do not have a clear purpose or goal. For socio-economic systems, on the other hand, this is more often the case. The purpose could be influenced by the contextual conditions given by market dynamics relative to a value creation potential, or by the formal or informal influence of institutional arrangements. Organizational capabilities are strategic resources with a function that contributes to competitive advantage (Barney 1991), hence, important for the organization and managed purposefully to co-create value with other intra- or inter-organizational capabilities (Helfat and Winter 2011). In business ecosystems, actors collaborate around a shared set of value propositions, constantly recombining complementarity of service exchange and resources (Cobben et al. 2022). In organizational routines actors perform their actions as an "effortfull accomplishment" as a contribution to uphold the patterning (Feldman et al. 2016).

Conclusion

Ultimately, this chapter demonstrates that any system across different levels of a nested hierarchy evolves through complexification processes in which it forms from a variety of elements related through interactive patterns. We focus on systems on a socio-economic level that are complex dissipative structures with nonlinear behavior, which also demonstrate an intent and ability to coordinate service exchange to co-create a shared set of value propositions. Through an interdisciplinary lens founded in a systemic-evolutionary perspective, we elucidate isomorphic principles that apply to socio-economic systems across different levels of a nested hierarchy. The principles support our definition of a socio-economic ecosystem as *a constraint on the variety of interactions and co-creation among agents and resources that maintains a shared set of focal value propositions in an institutional context.*

We demonstrate how any such socio-economic ecosystem emerges from dissipative organizing structures that apply constraints on variety. Further, how the system adapts and co-evolves from a process of complexification is demonstrated. This approach is generalizable to any type of organizing structure in a socio-economic realm and corresponds well with the categories found in organization theory and service theory. We hope it can contribute to understanding phenomena across multiple levels, where a holistic approach allows us to see beyond material elements, recognizing the properties that emerge from collective dynamics. In doing so, we gain a richer understanding of the processes that shape our social and economic landscapes, paving the way for more informed and effective strategies for managing complexity in an ever-changing world.

References

Adner, R.: Ecosystem as structure: an actionable construct for strategy. J. Manag. **43**(1), 39–58 (2017). https://doi.org/10.1177/0149206316678451

Anderson, P.: Complexity theory and organization science. Organ. Sci. **10**(3), 216–232 (1999)

Arthur, W.B.: Complexity and the economy. Science **284**(5411), 107–109 (1999)

Ashby, W.R.: Requisite variety and its implications for the control of complex systems. Cursos Congr. Univ. Santiago De Compostela **1**, 83–99 (1958)

Axelrod, R.: The Complexity of Cooperation: Agent-Based Models of Competition and Collaboration. Princeton University Press. https://www.jstor.org/stable/j.ctt7s951 (1997)

Barile, S., Polese, F.: Linking the viable system and many-to-many network approaches to service dominant logic and service science (SSRN Scholarly Paper 2003127). https://papers.ssrn.com/abstract=2003127 (2010)

Barney, J.: Firm resources and sustained competitive advantage. J. Manag. **17**(1), 99–120 (1991). https://doi.org/10.1177/014920639101700108

Bateson, G.: Steps to an Ecology of Mind: Collected Essays in Anthropology, Psychiatry, Evolution, and Epistemology. University of Chicago Press (1972)

Beer, S.: Brain of the Firm, 2nd edn. Wiley (1995)

Campbell, D.T.: Evolutionary epistemology. In: The philosophy of Karl R Popper, vol. 14–1, pp. 413–463. Open Court Publishing Company (1974)

Cobben, D., Ooms, W., Roijakkers, N., Radziwon, A.: Ecosystem types: a systematic review on boundaries and goals. J. Bus. Res. **142**, 138–164 (2022). https://doi.org/10.1016/j.jbusres.2021.12.046

Coleman, J.: Foundations of Social Theory. Belknap Press (1990)

Complex | Search Online Etymology Dictionary. (n.d.). Accessed July 11, 2024, from https://www.etymonline.com/search?q=complex

Darwin, C., Huxley, J.: The Origin of Species: 150th Anniversary Edition (Rep Anv edition). Signet (2003)

Dosi, G., Nelson, R., Winter, S.: The nature and dynamics of organizational capabilities. OUP Oxford (2001)

Feldman, M.S., Pentland, B.T.: Reconceptualizing organizational routines as a source of flexibility and change. Adm. Sci. q. **48**(1), 94–118 (2003). https://doi.org/10.2307/3556620

Feldman, M.S., Pentland, B.T., D'Adderio, L., Lazaric, N.: Beyond routines as things: introduction to the special issue on routine dynamics. Organ. Sci. **27**(3), 505–513. https://doi.org/10.1287/orsc.2016.1070 (2016)

Granovetter, M.: The strength of weak ties: a network theory revisited. Sociol Theory **1**, 201–233 (1983). https://doi.org/10.2307/202051

Gulati, R., Sytch, M., Tatarynowicz, A.: The rise and fall of small worlds: exploring the dynamics of social structure. Organ. Sci. **23**(2), 449–471 (2012). https://doi.org/10.1287/orsc.1100.0592

Gupta, A., Panagiotopoulos, P., Bowen, F.: Developing capabilities in smart city ecosystems: a multi-level approach. Organ. Stud. **44**(10), 1703–1724 (2023). https://doi.org/10.1177/01708406231164114

Hannan, M.T., Freeman, J.: The population ecology of organizations. Am. J. Sociol. **82**(5), 929–964 (1977). https://doi.org/10.1086/226424

Helfat, C.E., Winter, S.G.: Untangling dynamic and operational capabilities: strategy for the (N)ever-changing world. Strateg. Manag. J. **32**(11), 1243–1250 (2011). https://doi.org/10.1002/smj.955

Hernes, T.: Understanding Organization as Process: Theory for a Tangled World, 1st edn. Routledge (2007)

Heylighen, F.: (Meta)systems as constraints on variation—a classification and natural history of metasystem transitions. World Futur. **45**(1–4), 59–85 (1995). https://doi.org/10.1080/02604027.1995.9972554

Heylighen, F.: The growth of structural and functional complexity during evolution. In Evolution of Complexity, pp. 17–44. Kluwer Academic (1999)

Heylighen, F.: Self-organization of complex, intelligent systems: an action ontology for transdisciplinary integration. Integr. Rev. **134**(6), 1–39 (2011)

Heylighen, F., Cilliers, P., Gershenson, C.: Complexity and Philosophy (2006). http://arxiv.org/abs/cs/0604072

Holland, J.H.: Studying complex adaptive systems. J. Syst. Sci. Complex. **19**(1), 1–8 (2006). https://doi.org/10.1007/s11424-006-0001-z
Holland, J.H.: Signals and Boundaries: Building Blocks for Complex Adaptive Systems (Reprint edition). MIT Press (2014)
Jacobides, M.G., Cennamo, C., Gawer, A.: Towards a theory of ecosystems. Strateg. Manag. J. **39**(8), 2255–2276 (2018). https://doi.org/10.1002/smj.2904
Kauffman, S.: At Home in the Universe: The Search for the Laws of Self-Organization and Complexity (Revised edition). Oxford University Press (1996)
Klein, J.T.: Discourses of transdisciplinarity: looking back to the future. Futures **63**, 68–74 (2014). https://doi.org/10.1016/j.futures.2014.08.008
Koskela-Huotari, K., Svärd, K., Williams, H., Trischler, J., Wikström, F.: Drivers and hinderers of (un)sustainable service: a systems view. J. Serv. Res. **27**(1), 106–123 (2024). https://doi.org/10.1177/10946705231176071
Luhmann, N.: Social Systems. Stanford University Press (1995)
Lusch, R.F., Vargo, S.L.: Service-Dominant Logic: Premises, Perspectives. Cambridge University Press, Possibilities (2014)
Manderson, A.K.: A systems based framework to examine the multi-contextural application of the sustainability concept. Environ. Dev. Sustain. **8**(1), 85–97 (2006). https://doi.org/10.1007/s10668-005-2787-6
Maturana, H.R., Varela, F.J.: Autopoiesis and Cognition: The Realization of the Living. Springer Science and Business Media (1991)
Metcalfe, R.: Metcalfe's law after 40 years of ethernet. Computer **46**(12), 26–31 (2013)
Moore, J.F.: The Death of Competition: Leadership and Strategy in the Age of Business Ecosystems (Reprint edition). Harper Paperbacks (1997)
Nayak, A., Chia, R.: Thinking becoming and emergence: Process philosophy and organization studies. In: Tsoukas, H., Chia R., (eds.) Philosophy and Organization Theory, vol. 32, pp. 281–309. Emerald Group Publishing Limited (2011). https://doi.org/10.1108/S0733-558X(2011)0000032012
Nelson, R.R., Winter, S.G.: An Evolutionary Theory of Economic Change. An Imprint of Harvard University Press, Belknap Press (1985)
Parmigiani, A., Howard-Grenville, J.: Routines revisited: exploring the capabilities and practice perspectives. Acad. Manag. Ann. **5**(1), 413–453 (2011). https://doi.org/10.1080/19416520.2011.589143
Pentland, B.T., Hærem, T., Hillison, D.: Comparing organizational routines as recurrent patterns of action. Organ. Stud. **31**(7), 917–940 (2010). https://doi.org/10.1177/0170840610373200
Polese, F., Payne, A., Frow, P., Sarno, D., Nenonen, S.: Emergence and phase transitions in service ecosystems. J. Bus. Res. **127**, 25–34 (2021). https://doi.org/10.1016/j.jbusres.2020.11.067
Prigogine, I., Stengers, I.: Order out of Chaos. Bantam Books (1984)
Rescher, N.: Process philosophy: a survey of basic issues. University of Pittsburgh Press (2000)
Scott, W.R.: Institutions and Organizations: Ideas, Interests, and Identities. SAGE Publications (2013)
Simon, H.A.: The architecture of complexity. In: Proceedings of the American Philosophical Society vol. 106, pp. 467–482 (1962)
Storbacka, K., Nenonen, S.: Markets as configurations. Eur. J. Mark. **45**(1/2), 241–258 (2011). https://doi.org/10.1108/03090561111095685
System | Search Online Etymology Dictionary. (n.d.). Accessed July 11, 2024, from https://www.etymonline.com/search?q=system
Taillard, M., Peters, L.D., Pels, J., Mele, C.: The role of shared intentions in the emergence of service ecosystems. J. Bus. Res. **69**(8), 2972–2980 (2016). https://doi.org/10.1016/j.jbusres.2016.02.030
Teece, D.J., Pisano, G., Shuen, A.: Dynamic capabilities and strategic management. Strateg. Manag. J. **18**(7), 509–533 (1997)

Tuominen, T., Edvardsson, B., Reynoso, J.: Institutional change and routine dynamics in service ecosystems. J. Serv. Mark. **34**(4), 575–586 (2020). https://doi.org/10.1108/JSM-06-2019-0243
Vargo, S.L., Lusch, R.F.: Institutions and axioms: an extension and update of service-dominant logic. J. Acad. Mark. Sci. **44**(1), 5–23 (2016). https://doi.org/10.1007/s11747-015-0456-3
Vargo, S.L., Peters, L., Kjellberg, H., Koskela-Huotari, K., Nenonen, S., Polese, F., Sarno, D., Vaughan, C.: Emergence in marketing: an institutional and ecosystem framework. J. Acad. Mark. Sci. **51**(1), 2–22 (2023). https://doi.org/10.1007/s11747-022-00849-8
Von Bertalanffy, L.: The history and status of general systems theory. Acad. Manag. J. **15**(4), 407–426 (1972). https://doi.org/10.2307/255139
Weick, K.E., Sutcliffe, K.M., Obstfeld, D.: Organizing and the process of sensemaking. Organ. Sci. (2005). https://doi.org/10.1287/orsc.1050.0133
Winter, S.: Habit, deliberation, and action: strengthening the microfoundations of routines and capabilities. Acad. Manag. Perspect. **27**(2), 120–137 (2013)

Chapter 5
New Approaches to Steer Sustainability Transitions Within Service Ecosystems in the Context of Municipalities

Anna Viljakainen, Kirsi Maria Hyytinen, and Katri Valkokari

Abstract This chapter highlights the active role of municipalities in sustainability transitions. It adopts a service-ecosystems view in the public management domain to identify the systemic nature of impact and the dynamic role of the different actors in resource integration for increasing environmental, economic, cultural and social sustainability. It offers a new approach to steer the mindset change, give the direction and capture the long-term impacts of value co-creation, and systematically examine the interactions and dimensions of value at the micro-, meso-, and macro-levels in municipalities. The aim is to offer a more balanced and comprehensive picture of the dynamics of the municipal service systems to help identify predictable patterns for positive leverage points, and steer value co-creation in the complex systems towards positive sustainability impacts in the long-term. As a case-study with a design science approach, we co-created a framework with a Finnish municipality where the approach has been operationalized as part of strategy reorganization process. The outcome is a holistic and systemic framework that looks at the societal impacts in parallel with the techno-economic impacts, while unfolding the impacts on six dimensions: the residents, community, society, reputation, infrastructure and financial resources.

Keywords Municipality · Service ecosystems · Sustainability transitions · Value co-creation · Impact

A. Viljakainen
Aalto University School of Arts, Design and Architecture, Helsinki, Finland
e-mail: anna.viljakainen@aalto.fi

K. M. Hyytinen (✉) · K. Valkokari
VTT Technical Research Centre of Finland, Espoo, Finland
e-mail: kirsi.hyytinen@vtt.fi

K. Valkokari
e-mail: katri.valkokari@vtt.fi

D. Sarno et al. (eds.), *Complexity and Emergence in Market Ecosystems*, New Economic Windows, https://doi.org/10.1007/978-3-031-81942-1_5

Introduction

Over half of the world's population live in municipal areas, and the share is steadily growing in the coming decades (IPCC 2022). The rising city densities accelerate 'wicked' societal problems including climate change, public heath crises (Bettencourt and West 2010), loss of biodiversity, crime, societal fragmentation, and complexity (Di Meglio 2013). On the other hand, municipalities with increasing concentrations of wealth, people, power, and politics have become the centres of attention to fight against the sustainability challenges (Bettencourt and West 2010). Accordingly, they are seen as both the sources and the solution to the problems the world is facing. The complexity of social systems and the dynamic nature of sustainability (Loorbach 2010) put the networked municipalities in a key position to plan and execute strategies that contribute to sustainability transitions (STs) with large-scale and radical societal transformations involving institutional, social, cultural, organizational, and technological change (Köhler et al. 2019).

Such systemic changes demand for complex interaction between the various actors, resources, institutional practices, and regulations (Vargo, Wieland and Akaka 2015), and collaboration across public, private, and third sectors of society (Gallouj, Rubalcaba and Windrum 2013). A better understanding of the collaborative and highly complex interaction between decision makers, private and public actors, and citizens is highlighted (Windrum and Garcia-Goñi 2008). In this chapter municipalities are viewed as service ecosystems (e.g., Sarno et al. 2024) to provide a foundation for value co-creation within the variety of actors, including the municipality, its citizens and private sector organizations. With the theoretical lens of service-dominant logic that permits the analysis of markets (institutionalized solutions) as service ecosystems (Vargo et al. 2015), we focus on value cocreation between these interdependent actors in the municipal area (Osborne 2018; Osborne, Radnor and Nasi 2013). A service ecosystem is defined as "a relatively self-contained, self-adjusting system of resource-integrating actors connected by shared institutional arrangements and mutual value creation through service exchange" (Vargo and Lusch 2016, pp. 10–11). It highlights the role of institutions and social structures that guide interactions (Vargo et al. 2015) in a constantly adapting urban environment (Bettencourt and West 2010). In essence, the interactions between the various actors reform the institutions that form the markets (Vargo et al. 2015). Considering municipalities as service ecosystems allows us to identify the dynamic role of the different actors in other actors' value creation activities and the process of resource integration (Vargo and Lusch 2017) to support transformation of socio-technical systems at the macro-level (Ekman, Röndell and Yang 2019), mind-set change towards sustainability and resilience (Sebhatu and Enquist 2016) and de-/re-institutionalization (Friend et al. 2016; Sarno et al. 2024) at the meso-level, and societal wellbeing at the micro-level (Gallan et al. 2019).

Service-ecosystems view centres on the integration of dynamic resources to connect the social and technological facets of markets (Vargo et al. 2015), making it a useful theoretical lens to better understand and design policies, regulations, and funding structures that are necessary for STs in municipalities (e.g., Sebhatu, Johnson and Enquist 2016; Trischler, Johnson and Kristersson 2020). Yet, we experience a scarcity of empirical studies on value-based perspectives in the public service sector that require fundamental changes in managerial worldviews—shift in focus from linear co-production to dynamic value co-creation.

In the New Public Management (NPM) domain, new forms of non-hierarchical, de-centralized governance mechanisms have been demanded due to the increasing societal fragmentation, complexity and dynamism (Lévesque 2013). Along with the focus on STs, the post-NPM theories have emphasized the achievement of societal impact to indicate value generated in public sector management (Osborne et al. 2022). This has shifted the focus in the public management discourse from internal organizational efficiency and productivity, putting it rather into the fulfilment of societal values encompassing value on peoples' lives, environment and the society as a whole (Osborne, Nasi and Powell 2021).

Although the public and societal value and the practices to public management have been topical issues in recent discussion, the concept of "value" continues to be undertheorized and poorly understood both in theory and practice (Jo and Nabatchi 2016). Especially the empirical understanding how the value is defined and operationalised is limited (Osborne and Strokosch 2022). Therefore, a holistic framework has been sought to facilitate our understanding on the value co-creation in public services, contesting the common processes and mental models in city management previously locked-in to a traditional economics worldview (Osborne 2018). Service-ecosystems view yields a holistic, multilayered view on the (ST) process taking place in municipalities, being able to explain the reality of public administration that is fundamentally complex, interdependent, and fragmented by nature (Frow and Payne 2019; Osborne et al. 2013).

The purpose of this chapter is to understand multilayered dynamic value co-creation in municipalities' STs through a service-ecosystems view. In order to shed light on the new approaches for steering STs in the context of municipalities, we provide a case study with a design science approach. We aim to respond to the following research question: *How value co-creation can be steered within service ecosystems in the context of municipalities to advance mindset change and sustainability transitions?*

The chapter is structured as follows. After the introduction, we review the literature of service ecosystems and the public management domain. These form the two theoretical lenses for our study on the novel steering approaches in public management. Then we present shortly the research design for our case. Finally, we discuss findings and elaborate their theoretical and managerial implications, our research limitations, and offer directions for future studies.

Theoretical Lenses

The concept of public value enables a better understanding of broad and diverse social goals that are critical in steering STs in municipalities. As such, we revisit the concept of value by viewing municipalities as service ecosystems and evaluate the benefits for individuals, industries, and the society. This comprehensive perspective encompasses ecological, political, social, and cultural dimensions, surpassing individual self-interests.

The following sections provide the theoretical lenses into how ecosystems generate impact in municipalities by providing insight to municipalities as service ecosystems, reflecting the needs for novel steering approaches in public management.

Municipalities as Service Ecosystems

Municipalities are increasingly viewed as service ecosystems (e.g., Sarno et al. 2024). This perspective recognizes the interdependence of actors within the ecosystem, viewing them as resources that are both value creators and value beneficiaries (service-for-service; Lusch and Vargo 2006), and dissolves the traditional division between the municipality and its residents, including citizens and private sector organizations. The adoption of a service-ecosystems perspective allows city governments to actively drive change towards sustainability (Helkkula and Arnould 2022), ethical behaviour, and social responsibility, as opposed to simply adapt to uncontrollable changes in their operational environment (Vargo and Lusch 2008). It also offers a better understanding on how to provide such conditions that increase environmental, economic, and social sustainability and wellbeing rather than service provision alone (Vargo and Lusch 2008). For example, facilitate the emergence of social innovations by introducing new technologies, policies, infrastructures, roles, and behaviours to transform socio-technical systems (e.g., Ekman et al. 2019; Gallan et al. 2019; Trischler et al. 2020) towards sustainable and resilient municipalities (e.g., Sebhatu and Enquist 2016; Sebhatu et al. 2016). This is aligned with the core idea of systems thinking, i.e. how based on broad interactive collaboration, an ecosystem can generate more value by using the same input as individual actors alone.

An ecosystem brings together a variety of different actors. In the context of municipalities, the multilayered service ecosystem (Fig. 5.1) consists of collaborations and interactions at micro-, meso- and macro-levels (cf. Chandler and Vargo 2011a,b; Lusch and Vargo 2014). Collaboration is based on reciprocity: different types of actors, such as public sector actors, citizens, and companies of different sizes make their own knowledge, expertise and networks available to the ecosystems. The benefits and value derived from these ecosystems are multifaceted and may emerge over different time horizons, influencing various levels of interactions and dimensions of value (Chandler and Vargo 2011a,b). Therefore, the impacts of ecosystem collaboration and value co-creation emerge within the different levels and value dimensions.

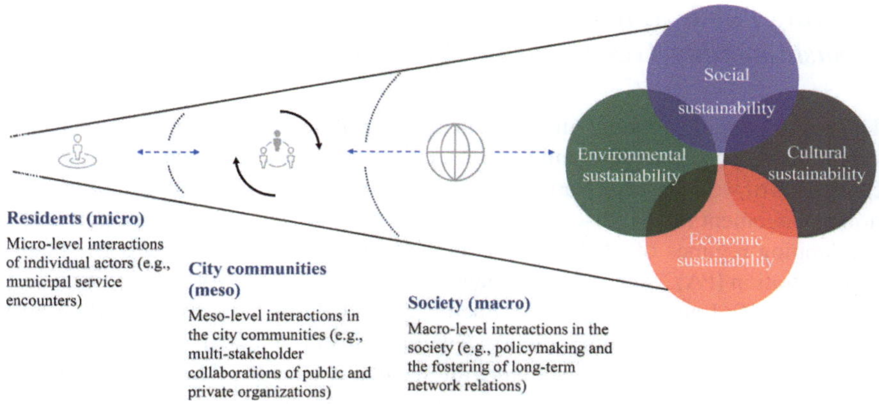

Fig. 5.1 Nested levels of interactions for creating long-term and multi-level impacts in the municipalities

These different levels are nested within one another that evolve and change over time (Vargo et al. 2015).

Through the continuous interaction of components (i.e., ecosystem actors and their resources), each ecosystem layer forms new properties: new resources, values, institutional arrangements and practices (Polese et al. 2021). In addition to internal dynamics, other layers and external environment may disturb or boost the interactions within the ecosystem. Hence, seeing value as a systemic property cannot be understood in isolation only focusing on one component or ecosystem layer (Akaka, Vargo and Lusch 2013). Studying value at the different levels of aggregation is necessary to fully comprehend the simultaneous micro- meso- and macro-level phenomena (Chandler and Vargo 2011a,b). In essence, system approaches are required to guide the development/impact building/value co-creation.

Fundamentally, the value creation relationships in municipalities are not dyadic firm-customer relationships but are profoundly dependent and influenced by the relationships and interactions between the many actors in the service ecosystems (actor-to-actor; Vargo and Lusch 2011). The operating environment can be seen as a resource to co-create a more sustainable (social, environmental, cultural, economic) future. For this reason, municipalities are increasingly focusing on providing such conditions that increase individuals' vitality and wellbeing. New tools and approaches are established for service co-creation and experimental learning to promote this goal. For example, the establishment of an amusement park near children's hospital to enhance the patient experience, which can also lead to positive impacts on their clinical outcomes (e.g., Martin et al. 2005). Therefore, integrative evaluation approaches are sought to identify predictable patterns on how markets behave and transform as complex systems in the long-term (Vargo et al. 2017).

Evolving Public Management: Addressing Needs of Sustainability Transitions in Municipalities

There is an ongoing change in the strategies of public management, which reconstructs its responses to economic and social crises, weakened social links and the challenges of welfare state (Harr et al. 2010; Osborne 2010). In the public administration management (PAM) discourse, researchers have identified three distinctive and overlapping regimes of public policy implementation and service delivery: public administration (PA), New Public Management (NPM) and network governance (e.g. Osborne 2010; Langergaard 2011). Each regime contains a different 'world view': particular assumptions about societal needs and a diverse understanding of the means and roles of different actors in tackling the societal challenges. They reflect a further change in the focus of public service provision: from bureaucratic and rule-based order to the emphasis on cost-efficiency, markets and performance and in the latter phase relationships, partnerships and co-creation (Torfing and Triantafillou 2013).

In the recent discourse, the current management regimes have proved incompatible with the current societal developments (Osborne 2018, 2021). Root cause is that the management theory is deriving from goods producing firms and ignoring the fact that the public sector is devoted to generating a platform for co-creation with citizens, industry and decision makers in a sustainable manner (Bryson et al. 2017). While earlier regimes still have a strong position in the public sector management and the different elements of each regime typically coexist and operate in the shadow of the most traditional ones, there is a new paradigm emerging that aims to respond to the current societal trends such as sustainability, increasing choice of citizens, inclusiveness, non-hierarchical and self-regulated practices and co-creation of value (Osborne 2021). The recent discourse on public administration and management theory has been influenced by service-dominant logic (Vargo and Lusch 2004, 2008, 2016) and introduced a new approach of Public Service Logic (PSL; Osborne, 2018, 2021). With the rise of PSL, the focus has increasingly shifted on the conceptualisation of value (Cui and Osborne 2021; Osborne et al. 2021) to better understand the management of public organisations treated as public service organisations (PSO; Engen et al. 2021).

Value has been a consistent element in the public management domain for decades. However, its critics have argued that the approach only focuses on monetary value including economic efficiency and productivity due to the NPM tradition which emphasizes the competitive state, efficiency in terms of the economy and market selection, and the replacement of public services by private actors (Lévesque 2013).

Increasing societal fragmentation, complexity, and dynamism have created a demand for non-hierarchical, decentralized governance mechanisms (Lévesque 2013). Post-NPM theories, with a focus on sustainability, emphasize achieving societal impact as a measure of value in public sector management (Osborne et al. 2022). This shift in public management discourse moves the focus from internal organizational efficiency and productivity to fulfilling societal values, enhancing people's lives, and benefiting the environment and society as a whole (Osborne et al. 2021,

2022). In defining value, emphasis has been placed on both individual citizens and the broader society, extending beyond self-interests to include ecological, political, social, and cultural dimensions (Benington 2009).

While networked forms of governance have garnered significant interest there is no consensus on whether they are becoming the dominant steering mechanism in the public sector. Some phenomena suggest a paradigmatic shift, while others indicate a merging of different paradigms. Currently, public steering and decision-making remain largely based on bureaucratic administration or NPM (Levesqué 2013; Moore and Hartley 2008). This focus on economic and technocratic views of services overlooks their collaborative, interactive, and dynamic nature, as well as sustainability and societal value as primary policy goals.

Consequently, public managers and policymakers need to prioritize the co-creation of societal value in public management. This requires a shift in focus from internal efficiency and productivity to external values and societal impact throughout public organizations (Osborne et al. 2022).

Approaches to Steer Sustainability Transitions in Municipalities

> "Cities are complex systems whose infrastructural, economic and social components are strongly interrelated and therefore difficult to understand in isolation. The many problems associated with urban growth and global sustainability, however, are typically treated as independent issues. This frequently results in ineffective policy and often leads to unfortunate and sometimes disastrous unintended consequences".

(Bettencourt and West 2010, pp. 912).

We experience a perspective shift in conceptualizing the municipal markets. A traditional focus has been consistent to a rather static (neoclassical) economics world-view, where the units of output (e.g., social/health/housing/education services and municipal infrastructure) have been treated as key components of exchange, and customers as targets for exchange (cf. Vargo and Lusch 2004). Value has been assigned to municipal services that are distributed to eligible residents as statutory duties. Primacy has been placed on techno-economic targets with the specific emphasis on related measures (Table 5.1) that focus on productivity and cost-efficiency. These targets however ignore the broader societal effects and phenomena that are meaningful from the perspective of broader society and citizens—their health, wellbeing, vitality, and competences, among others. Furthermore, a techno-economic approach does not consider impacts that emerge in dynamic and long-term processes interlinked with broader societal targets, structures and developments (Djellal and Gallouj 2010, 2013; Hyytinen, Saari and Elg 2019). Similarly, they ignore the systemic nature of value—before value can be realized, resources must be integrated by joining different actors into mutual value co-creation efforts (Vargo and Lusch 2011).

Table 5.1 Theoretical underpinnings for steering STs in municipalities

Traditional approach	New approach
Techno-economic	Societal
Approaches in steering one-sidedly technological and financial, without capturing the multiplicity and social and systemic nature of impact (Rubalcaba, Di Meglio and Gallego 2013; Djellal and Gallouj 2013). Economic activity is based on the production and distribution of products and services (transactions) that can be quantifiable in terms of exchange value, with primacy on maximizing service efficiency and profits (Vargo and Lusch 2004)	A multicriteria approach integrating techno-economic and broader societal targets (Djellal and Gallouj 2013) and different levels of aggregation (Chandler and Vargo 2011a, b) for a balanced and comprehensive picture of sustainability impacts. Seeing market being shaped via activities of social and economic actors and formed as socio-technical and systemic processes (Azimont and Araujo 2007). Focus on the collaboration between multiple actors from different sectors of society (Windrum and Garcia-Goñi 2008) and the systemic nature of value co-creation—before value can be realised, resources must be integrated (Vargo and Lusch 2011)
Atomistic	Systemic and holistic
Perceived regularity of markets; A mechanistic, reductionistic and/or atomistic worldview of the different parts of the systems with focus on components, structures, and countable objects (Vargo et al. 2017). Adoption of a dyadic value chain view separating the provider and the market (Vargo and Lusch 2004)	Markets are formed over time; A holistic, organismic and/or ecological worldview on the whole system with focus on long-term processes, relationships and interactions (Vargo et al. 2017). Perception of dynamic systems that adapt to changing conditions via transformation (Spohrer et al. 2007) that can be triggered by endogenous (e.g., internal interactions, innovations) and exogeneous causes (e.g., external shocks, disturbances; Koskela-Huotari et al. 2021) characterized by the emergence of new properties (e.g., resources, value, practices; Polese et al. 2021). Steering and target setting therefore interlinked to broader societal targets, structures and developments in dynamic, emergent, evolving, and adapting contexts (Osborne et al. 2022)

Source Authors' elaboration

Societal approach puts emphasis on integration of techno-economic and broader societal targets (Djellal and Gallouj 2013) and different levels of aggregation (Chandler and Vargo 2011a,b) for a balanced and comprehensive picture of sustainability impacts. The systemic approach therefore highlights the dynamic nature of value co-creation in service ecosystems, emphasizing knowledge and skills as the primary resources for exchange. By doing so, it shifts the focus on the activities and processes that feed value co-creation, rather than the outputs and outcomes that follow (Vargo and Lusch 2004, 2008). This value co-creation process is dynamic and value itself is multifaceted, i.e. different actors experience/capture value at different time horizon and in many different forms from individual well-being and cost efficiency to system-level sustainability.

Seeing municipalities as complex, self-organizing and adaptive systems necessitates the identification of predictable patterns (fractals) on relationships taking place among different levels of aggregation to unfold how systems behave and transform over time (Vargo et al. 2017). In the following chapter we attempt to concretize these relationships to steer the interactions in co-creating long-term value and impacts. By value, we refer to the foundational element of service (i.e., the application of knowledge and competences for the benefit of others) that unfolds in the interactions of different actors and profoundly affects the dynamics of service systems (Vargo, Maglio and Akaka 2008). The adoption of a service-ecosystems view shifts focus from the producer and transactions (value-in-exchange) to the perspective of the system and the use context in the determination of value: i.e., how value is co-created in interactions of providers and beneficiaries by integrating resources in specific contexts (value-in-use; Vargo and Lusch 2004, 2008). Impact, too, in service ecosystems centres around institutional arrangements, relationships and perceptions of value, rather than outputs (i.e., direct products) of pre-specified inputs (e.g., financial resources). It refers to system level outcomes (institutional and societal change) to be expected in the long-term (Djellal and Gallouj 2013).

Research Method

A design science approach suggests that research is about changing issues and phenomena, not just understanding and explaining them (Burke 2002). This study aimed to develop an artifact (March and Smith 1995) to help researchers and practitioners better grasp the dynamics of municipal service ecosystems. Design science can be linked to several qualitative methods like action research, participatory case studies, academia–industry partnerships, and constructive approaches. With the lens of design science, these exploratory approaches collectively aim for creating an artifact to address a practical problem and a 'means to an end' (Holmström et al. 2009). Our study's managerial challenge and research question is: "*How value co-creation can be steered within service ecosystems in the context of municipalities to advance mindset change and sustainability transitions*?".

A joint process of researchers and practitioners from the city, along with close collaboration during problem-solving, was essential in this research setting. In other words, through the case, both practitioners and researchers endeavoured to collectively create and assess the design science research artefact—a framework for impacts co-created in service ecosystems (Fig. 5.2). It is an outcome of a research and development project, where the authors were involved in strategy reorganization processes of a Finnish municipality located in Southern Finland. The design science process was a 12-week co-creation project carried out in fall 2020 and spring 2021. This project focused on the strategy level interaction in the municipality. Participants from the case organization were municipality officials actively involved in the strategy reorganization process. Democratically elected municipal councillors were not involved in the research process, because municipal elections in Finland were scheduled for spring 2021. Therefore, municipality officials decided to include new democratically elected councillors only at a later stage of the project. Data was gathered in workshops and online-meetings (altogether 22 people attending, some more than once), face-to-face semi-structured interviews (5 interviews recorded and transcribed), and from a large pool of publicly available municipal documents. The method was chosen for studying strategic phenomena, allowing interviewees to reflect on their day-to-day practices (Eisenhardt and Graebner 2007).

This conceptual artefact represents the primary outcome of the study. Design science methods entail a series of steps and an instantiation that operationalize these constructs, models, and methods —essentially, how they represent the jointly developed artefact in its environment. These key steps of the joint research process are outlined in Table 5.2.

The holistic and systemic framework for the municipality's sustainability impacts was built in a co-creation project to which both city officials and researchers participated. Corresponding frameworks have previously been developed to evaluate systemic innovation (Hyytinen 2017), sustainable service innovation (Hyytinen,

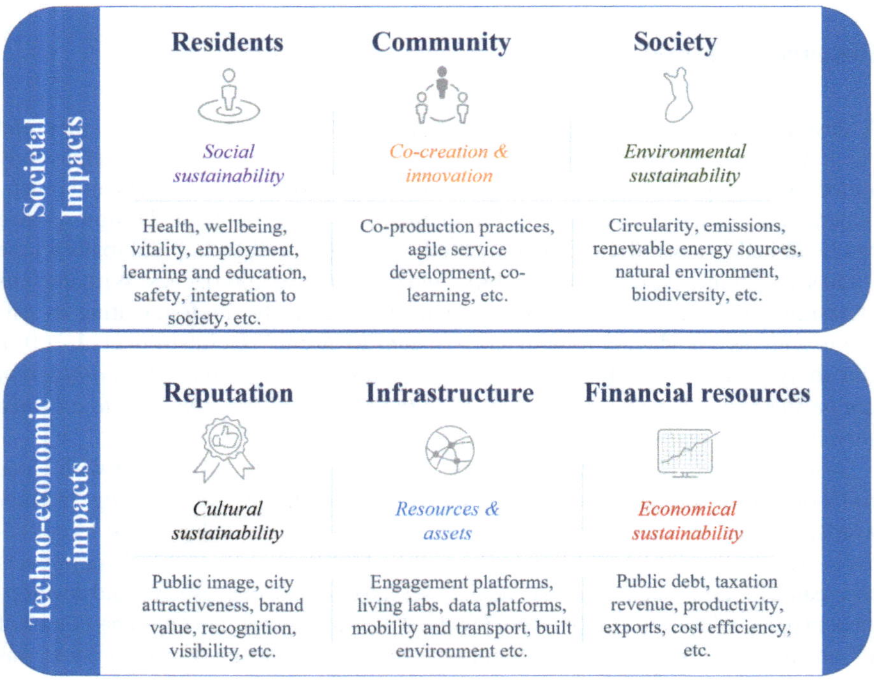

Fig. 5.2 Framework for impacts co-created in service ecosystems. *Source* Authors' elaboration

Table 5.2 Summary of the key research activities in the design science process of this study

	Build	Evaluate	Theorize/Justify
Constructs	Multi criteria evaluation Multi-agent approach Multi-level perspective on service ecosystems Value co-creation	Current impact-related target setting and measurement in the city bases on techno-economic approach on productivity Atomistic and diversified approach to value and impact Needs for novel comprehensive targets and indicators to enable STs in the city	Developing new comprehensive targets and related indicators with emphasis on both techno-economic and societal impacts to steer the city's STs Systemic approach to value and value co-creation Requirements and development needs in organizational and data management to enable STs
Model	Strategic sustainability targets of the city: Social, environmental, economic, and cultural sustainability Interactions at the micro-, meso- and macro-levels	Societal impacts: Social sustainability (residents), co-creation and innovation (community), environmental sustainability (society) Techno-economic impacts: cultural sustainability (reputation), resources and assets (infrastructure), economic sustainability (financial resources)	Framework presented in this chapter
Method	5 workshops in Dec 2020—Jun 2021 (à 90–120 min) 5 interviews with 4 interviewees in Jan 2021—Mar 2021 (à 60 min)	1st workshop: Long-term vision for the municipality in the ST 2nd workshop: Role of the city organization to steer the ST	3rd workshop: Framework and comprehensive impact targets for the social, cultural, economic and environmental sustainability 4th workshop: Impact indicators for each sustainability dimension 5th workshop: Roadmap with short-, mid- and long-term development activities and ST enablers around (1) Organizational management, (2) data management, and (3) co-creation in service ecosystems

Source Authors' elaboration

Saari and Elg 2019) and to support transformative sustainability management in cities (Bhatia et al. 2024). These frameworks have adopted the idea of multi-criteria evaluation approaches for innovation (see Djellal and Gallouj 2013). The studies have evaluated the social impacts of innovation, beyond the traditional techno-economical views, by investigating different justifications and conflicts between them. In this project, previous frameworks served as inspiration, on the basis of which they were further developed and applied in the context of service ecosystems for STs.

Operationalization of Novel Steering Approaches to Municipal Strategy: A Framework of Impacts Co-Created in Service Ecosystem

The new approach makes visible the interconnectedness and systemic properties of phenomena in complex systems (Vargo et al. 2017), to give direction and to allow the operationalization of strategy and related mindset change in municipalities. For the purpose of concretizing the new approach into managerial efforts in municipalities' steering, we propose a framework (Fig. 5.2) to crystallise societal and systemic value and impacts co-created in service ecosystems. As described at the Sect. 3, the framework has been further co-developed with practitioners based on a previously published model, which was designed to support the co-evaluation of human-centred experiments (Hyytinen et al. 2019).

The aim of this framework is to offer a more balanced and comprehensive analysis on the impacts co-created in municipalities by moving focus towards a service-ecosystems view, and adopting a strong sense of social responsibility, ethical and sustainable behaviour (Djellal and Gallouj 2010, 2013; Vargo and Lusch 2008). We recognize that the perspectives on impacts are fundamentally intertwined in our framework. For example, in the context of healthcare, patients' interactions in their individual service ecosystems at the hospital surroundings may increase patients' lived experience (i.e., value), leading to enhanced patient health and wellbeing (i.e., impact; see e.g., Gallan et al. 2019; Martin et al. 2005). These practices can be adopted into new roles, co-creation activities and behaviour in the healthcare service delivery, leading to increased social and economic sustainability not only at the micro-level (e.g., individual's prolonged life expectancy), but also at the meso-level (e.g., shortened time spent in hospitals) and macro-levels (e.g., growth in public health).

Accordingly, the framework offers a multi-dimensional approach by looking at the societal impacts in parallel with the techno-economic characteristics of innovations (Djellal and Gallouj 2010, 2013), while considering the shared institutional arrangements among actors at the different levels of aggregation (multi-level perspective; Chandler and Vargo 2011a, b). The dimension of societal impact separates between individual residents, the municipal community, and the society, suggesting that value is co-created in interaction at the micro-, meso-, and macro-levels (Vargo, Wieland and Akaka 2015). Here, impact targets may submit either qualitative or quantitative measures, and in the context of individual residents may refer to such indicators as (perceived) health, wellbeing, or vitality (*social sustainability*). This is where we recognise that value is always co-created and determined uniquely and phenomenologically by the beneficiary (value-in-context; Vargo and Lusch 2008), with the service provider having the aim to create successful customer experiences in the long-term. At the meso-level, our framework introduces the municipal community (i.e., residents and the public/private/third party organizations) as the dimension that focuses on *value co-creation and the fostering of innovation.* For example, impact indicators may focus on the emergence and evolution of concrete co-production practices, interactions, and resource mobilization within service ecosystems that drive value co-creation and re-invent the ways in which value is created in the actual service processes where the relationships are active (see also Gallan et al. 2019; Sebhatu et al. 2016). The last element of societal impacts is the macro level focusing on *ecological sustainability*. For example, tracking the developments and diffusion of user innovations and measures that promote environmental sustainability and carbon-free societies (see also Trischler, Johnson and Kristersson 2020).

Transformation towards a systemic view calls for a multidimensional approach that also considers the issues of quality and reputation (Djellal and Gallouj 2013). Actors in service ecosystems are dynamic resources that co-create value and co-produce the service as part of their network influenced by the institutions (Vargo et al. 2015). Therefore, brands are co-created in relationships and interactions among ecosystem actors (Merz, He and Vargo 2009), rather than built by service providers. Hence, our framework introduces *cultural sustainability* as a dimension of impact. Here, we emphasize the cultural diversity and sustainability, public image, and perceived city attractiveness being co-created in the service ecosystems.

The framework also introduces the actual *resources, assets* (e.g., platforms, digital technology, infrastructure), processes, and practices necessary for value co-creation and increased connectivity, which can promote and explicate the micro- and meso-level linkages in wellbeing (see also Reypens, Lievens and Blazevik 2016). Engagement platforms (Ramaswamy 2011) are means that facilitate the co-creation of value by allowing ongoing interactions between different actors in service ecosystems (e.g., physical co-creation spaces, digital data platforms, solutions that enhance mobility, etc.).

As a final dimension our framework introduces the *financial resources* as an impact target, with focus on the positive impacts from increases in social sustainability, collaboration, and value co-creation to municipal performance. Here, we emphasize economical sustainability and highlight a long-term orientation where the focus is on increasing efficiency through effectiveness instead of making efficiency primary (Vargo 2009).

Conclusion and Discussion

A new managerial mindset is being adopted, where the traditional view that municipalities unilaterally create value for citizen is replaced with the idea that value is co-created within systems of multi-actor networks. A shift of this magnitude requires fundamental changes in the mindsets to understand the dynamic service ecosystems, impact targets and related metrics to build transformational action for shaping resilient municipalities for the people, while supporting economic, societal, cultural and environmental wellbeing. In essence, the traditional techno-economic measures do not properly capture the diversity of the impacts that emerge in dynamic and long-term processes (Djellal and Gallouj 2010, 2013) in municipalities. Furthermore, biased measures may cause inaccurate analysis and interpretations and lead to inappropriate policies (Bettencourt and West 2010; Rubalcaba 2006). Therefore, organizations that participate in multi-stakeholder collaborations should have a clear understanding of the types of value outcomes that can be created and captured (Reypens et al. 2016). In the recent post NPM discourse the frameworks have been developed to analyse the ecosystemic approach for value creation in the public service context (e.g. Trichler et al. 2023; Osborne et al. 2021). The approaches are however theoretical and lack empirical evidence from municipal contexts.

We propose a holistic empirically build framework that allows municipalities to concretize the sustainability targets and systemize actions to adapt into new operations and mindsets. The approach provides an alternative by emphasizing the societal, systemic and collaborative nature of value co-creation in service ecosystems in the context of municipalities. It integrates a multi-dimensional framework to evaluate multiple impacts and a service-ecosystems view to crystallise the value from the perspective of multiple actors. The framework unfolds impacts of innovations on six dimensions. Specific emphasis is put on societal impacts, which are analysed in parallel with the traditional techno-economic impacts. Dimensions included are impacts on residents, community and society as well as impacts on reputation, infrastructure and financial resources.

A novel sustainability-oriented impact is essential for steering effective STs. While new policy tools and practices are often introduced to address sustainability challenges, their impact is limited if they are simply added to existing frameworks without being fully integrated into the governance and management systems (Jäntti et al. 2023). Moreover, while New Public Management emphasizes the need for governments to achieve ‘organizational sustainability,’ focusing too narrowly on the concept of value overlooking the long-term environmental and social consequences of public actions (Osborne et al. 2015). Therefore, a more strategic and comprehensive approach is required—embedding sustainability into the core of city operations to guide decision-making across all policy domains.

Implications and Future Research Avenues

Service-ecosystems view can act as a powerful managerial tool and lens in the formulation of a new vision for municipalities. When applied with a new steering approach and related impact targets, it yields great leverage in the reorganization and implementation of sustainable strategy. We observe that when a value-based steering approach is adopted as part of strategic management practices, a number of benefits take place. Firstly, adopting a comprehensive and systemic impact approach as part of strategy work allows municipalities to co-create a sustainable and resilient future together with others, rather that focus on continuously adapting to an uncontrollable environment and cost-cutting, because it helps to concretize the direction and pace of transformation as a co-creational process for all stakeholders. Secondly, adopting impact targets in operating planning promotes cultural transformation, because it systemizes organizational goals and engages key stakeholders in the process. Thirdly, taking a multi-perspective and multi-actor view allows municipalities to identify key competence gaps and measures to overcome them. In particular, it helps to break free from siloes by making systemic impacts visible. Fourthly, it allows municipalities to identify predictable patterns and verify the impacts of decisions and actions, when combined with qualitative and quantitative datasets. It also helps to uncover investment areas with the greatest leverage. And finally, it allows municipalities in business model innovation by making the connections between efficiency, effectiveness and impact transparent.

The complexity of societal problems demand for flexible and learning-oriented institutions, which contradicts the traditional emphasis on sector-specific hierarchies, rules, efficiencies, and accountability (Friend et al. 2016) in the public management. Dynamic evaluation systems constantly change because of collaborative learning and self-reflection (Hyytinen et al. 2019). Therefore, a longitudinal study is recommended to detect changes how the new approach to evaluation and impact targets capture long-term impacts of value co-creation, collaborative learning, and related mindset change in municipalities. In specific, we recommend studying emergent patterns in the processes of instilling a value-based mindset in operating planning by articulating strategic goals into concrete Key Performance Indicators (KPIs) and action to lead mindset change and transition at the micro-level. On the other hand, special attention should be paid on the processes of institutionalization of the new rules and practices at the meso- and macro-levels. Here we encourage the mapping of repeat and predictable patterns (fractals) in longitudinal data on the relationships taking place among different aggregation levels to unfold system level change over time (Vargo et al. 2017). Having a systems perspective will better allow scholars to detect how the micro-level phenomena may initiate and even lead meso- and macro-level phenomena. Additionally, our analysis, which is grounded in managerial perspectives, does not account for insights from politicians and sectoral experts. Future research should explore the implications of applying a managerial lens to inherently political issues such as sustainable development, particularly in relation to its effects on democratic processes, policymaking, and citizen engagement.

References

Akaka, M.A., Vargo, S.L., Lusch, R.F.: The complexity of context: a service ecosystem approach for international marketing. J. Int. Mark. **21**(4), 1–20 (2013)

Azimont, F., Araujo, L.: Category reviews as market-shaping events. Ind. Mark. Manage. **36**(7), 849–860 (2007)

Benington, J.: Creating the public in order to create public value? Int. J. Public Adm. **32**(3–4), 232–249 (2009)

Bettencourt, L., West, G.: A unified theory of urban living. Nature **467**, 912–913 (2010)
Bhatia, R., Pihlajamaa, M., Hyytinen, K., Jäntti, A.: Capabilities for transformative sustainability management in cities. Public Manag. Rev. 1–25 (2024)
Bryson, J., Sancino, A., Benington, J., Sørensen, E.: Towards a multi-actor theory of public value co-creation. Public Manag. Rev. **19**(5), 640–654 (2017)
Burke, W.: Organization Change: Theory and Practice. SAGE Publications, Thousand Oaks, California (2002)
Chandler, J.D., Vargo, S.L.: Contextualization and value-in-context: how context frames exchange. Mark. Theory **11**(1), 35–49 (2011a)
Chandler, J.D., Vargo, S.L.: Contextualization: Network intersections, value in-context, and the cocreation of markets. Mark. Theory **1**, 35–49 (2011b)
Cui, T., Osborne, S.P.: Unpacking value destruction at the intersection between public and private value. *Public Adm.* 1–20 (2021)
Di Meglio, G.: The Place of ServPPINs in the range of public-private collaboration arrangements for service provision. In: Gallouj, F., Rubalcaba, L., Windrum, P. (eds.) Public-private innovation networks in services. Edgar Elgar, Northampton, Cheltenham, pp. 59–87 (2013)
Djellal, F., Gallouj, F.: The Innovation gap and the performance gap in the service economies: a problem for public policy. In: Gallouj, F., Djellal, F. (eds.) The Handbook of Innovation in Services. A Multi-disciplinary Perspective. Edgar Elgar, Northampton, Cheltenham, pp. 653–673 (2010)
Djellal, F., Gallouj, F.: The Productivity in services: measurement and strategic perspectives. Serv. Ind. J. **33**(3–4), 282–299 (2013)
Eisenhardt, K.M., Graebner, M.E.: Theory building from cases: opportunities and challenges. Acad. Manag. J. **50**(1), 25–32 (2007)
Ekman, P., Röndell, J., Yang, Y.: Exploring smart cities and market transformations from a service-dominant logic perspective. Sustain. Cities Soc. **51**, 1–11 (2019)
Engen, M., Fransson, M., Quist, J., Skålén, P.: Continuing the development of the public service logic: a study of value co-destruction in public services. Public Manag. Rev. **23**(6), 886–905(2021)
Friend, RM., et al.: Re-imagining inclusive urban futures for transformation. Curr. Opin. Environ. Sustain. 20, 67–72 (2016)
Frow, P., Payne, A.: Value Cocreation: An Ecosystem Perspective. In: Vargo, S.L., Lusch, R.F. (eds.) The SAGE handbook of service-dominant logic, pp. 80–96. Sage Publications, London, UK (2019)
Sarno, D., Enquist, B., Polese, F., Sebastiani, F., Sebhatu, S., Viljakainen, A.: A processual view on sustainability transitions in service ecosystems. J. Serv. Manag. (ahead-of-print) (2024)
Gallan, A.S., et al.: Transforming community well-being through patients lived experiences. J. Bus. Res. **100**, 376–391 (2019)
Gallouj, F., Rubalcaba, L., Windrum, P. (eds.): Public-private innovation networks in services. Edward Elgar, Cheltenham and Northampton (2013)
Harrisson, D., Klein, J.-L., Browne, P.L.: Social innovation, social enterprise and services. In: Gallouj, F., Djellal, F. (eds.) The handbook of innovation and services, pp. 197–281. Edward Elgar, Cheltenham and Northampton (2010)
Helkkula, A., Arnould, E.J.: Using neo-animism to revisit actors for sustainable development goals (SDG's) in S-D logic. J. Bus. Res. **149**, 860–868 (2022)
Holmström, J., Ketokivi, M., Hameri, A.-P.: Bridging practice and theory: a design science approach. Decis. Sci. **40**(1), 1540–5915 (2009)
Hyytinen, K, Saari, E., Elg, M.: Human-Centered Co-Evaluation method as a means for sustainable service innovations. In: Toivonen, M., Saari, E. (eds.) Human-Centered Digitalization and Services. Springer, Singapore (2019)
Hyytinen, K.: Supporting service innovation via evaluation: a future oriented, systemic and multi-actor approach, Doctoral dissertation 14/2017, Aalto University publication series and VTT Science 146 (2017)

IPCC (2022), "Cities, Settlements and Key Infrastructure", Delgado-Ramos, D.C., and Romero-Lankao, P. (Eds.), *Climate Change 2022: Impacts, Adaptation and Vulnerability. Contribution of Working Group II to the Sixth Assessment Report of the Intergovernmental Panel on Climate Change*, Cambridge University Press, Cambridge, UK and New York, NY, pp. 907–1040.

Jo, S., Nabatchi, T.: Getting back to basics: advancing the study and practice of coproduction. Int. J. Public Adm. **36**(13), 1101–1108 (2016)

Jäntti, A., Paananen, A., Kurkela, K.: Towards interactive governance: embedding citizen participation in local government. Administration & Society **55**(8), 1529–1554 (2023). https://doi.org/10.1177/00953997231177220

Koskela-Huotari, K., et al.: Service system transformation through service design: linking analytical dimensions and service design approaches. J. Bus. Res. **136**, 343–355 (2021)

Köhler, J., et al.: An agenda for sustainability transitions research: state of the art and future directions. Environ. Innov. Soc. Trans. **31**, 1–32 (2019)

Langergaard, L.: Understanding of users and innovation in a public sector context. In: Sunbo, J., Toivonen, M. (eds.) User-based Innovation in Service, pp. 203–226. Edwards Elgar, Cheltenham (2011)

Lévesque, B.: Social Innovation in governance and public management systems: toward a new paradigm? In: Moulaert, F., MacCallum, D., Mehmood, A., Hamdouch, A. (eds.) The International Handbook of Social Innovation. Collective Action, Social Learning and Transdisciplinary Research, pp. 25–39. Edward Elgar, Cheltenham (2013)

Loorbach, D.: Transition management for sustainable development: a prescriptive, complexity-based governance framework. Governance **23**(1), 161–183 (2010)

Lusch, R.F., Vargo, S.L.: Service-dominant logic as a foundation for a general theory. In: Lusch, R.F., Vargo, S.L. (eds.) The service-dominant logic of marketing: Dialog, debate, and directions, pp. 406–421. Armonk, NY: M.E. Sharpe (2006)

Lusch, R.F., Vargo, S.L.: The Service-Dominant logic of marketing: dialog, debate, and directions. Routledge, London, UK (2014)

March, S.T., Smith, G.F.: Design and natural science research on information technology. Decis. Support. Syst. **15**(4), 251–266 (1995)

Martin, L., Williams, S., Haskard, K., Dimatteo, M.: The challenge of patient adherence. Ther. Clin. Risk Manag. **1**(3), 189–199 (2005)

Merz, M.A., He, Y., Vargo, S.L.: The evolving brand logic: a service-dominant logic perspective. J. Acad. Mark. Sci. **37**, 328–344 (2009)

Moore, M., Hartley, J.: Innovations in Governance. Public Manag. Rev. **10**(1), 3–20 (2008)

Osborne, S.P.: The New Public Governance? Routledge, London, UK (2010)

Osborne, S.P.: From public service-dominant logic to public service logic: are public service organizations capable of co-production and value co-creation? Public Manag. Rev. **20**(2), 225–231 (2018)

Osborne, S.P.: Public service logic. Creating value for public service users, citizens, and society through public service delivery. New York, NY: Routledge (2021)

Osborne, S., Strokosch, K.: Participation: add-on or core component of public service delivery? Aust. J. Public Adm. **81**, 181–200 (2022)

Osborne, S.P., Radnor, Z., Nasi, G.: A new theory for public service management? toward a (public) service-dominant approach. Am. Rev. Public Adm. **43**(2), 135–158 (2013)

Osborne, S.P., Nasi, G., Powell, M.: Beyond co-production: value creation and public services. Public Administration **99**, 641–657 (2021)

Osborne, S., Powell, M., Cui, T., Strokosch, K.: Value creation in public service ecosystem: an integrative framework. Public Adm. Rev. **82**(4), 634–645 (2022)

Polese, F., et al.: Emergence and phase transitions in service ecosystems. J. Bus. Res. **127**, 25–34 (2021)

Ramaswamy, V.: It's about human experience and beyond, to co-creation. Ind. Mark. Manage. **40**(2), 195–196 (2011)

Reypens, C., Lievens, A., Blazevik, V.: Leveraging value in multi-stakeholder innovation networks: a process framework for value co-creation and capture. Ind. Mark. Manage. **56**, 40–50 (2016)

Rubalcaba, L., Di Meglio, G., Gallego, J.: Public-private innovation networks and social innovation in the service economy. In: Vinals, C.R., Rodriguez, C.P. (eds.) Social innovation: new forms of organization in knowledge-based societies, pp. 188–205. Routledge, New York, USA (2013)

Sebhatu, S.P., Enquist, B.: Sustainable public transit service value network for building living cities in emerging economies: multiple case studies from public transit services. Procedia–Soc. Behav. Sci. 22, 4263–268 (2016)

Sebhatu, S.P., Johnson, M., Enquist, B.: Service innovation in complex service system: public transit service sustainability business cases. Procedia–Soc. Behav. Sci. 224, 269–275 (2016)

Spohrer, J., Maglio, P.P., Bailey, J., Gruhl, D.: Steps toward a science of service systems. Computer **40**(1), 71–77 (2007)

Torfing, J., Triantafillou, P.: What's in a name? Grasping new public governance as a political-administrative system. Int. Rev. Public Adm. **18**(2), 9–25 (2013)

Trischler, J., Røhnebæk, M., Edvardsson, B., Tronvoll, B.: Advancing public service logic: moving towards an ecosystemic framework for value creation in the public service context. Public Manag. Rev. 1–29 (2023)

Trischler, J., Johnson, M., Kristensson, P.: A service innovation perspective on the diffusion of sustainability-oriented user innovations. J. Bus. Res. **116**, 552–560 (2020)

Vargo, S.L.: Toward a transcending conceptualization of relationship: a service-dominant logic perspective. J. Bus. Ind. Mark. **24**(5/6), 373–379 (2009)

Vargo, S.L., Lusch, R.F.: Evolving to a new dominant logic. J. Mark. **68**(1), 1–17 (2004)

Vargo, S.L., Lusch, R.F.: From goods to service(s): divergences and convergences of logics. Ind. Mark. Manage. **37**(3), 254–259 (2008)

Vargo, S.L., Lusch, R.F.: It's all B2 and beyond: toward a system perspective of the market. Ind. Mark. Manage. **40**(2), 181–187 (2011)

Vargo, S.L., Lusch, R.F.: Institutions and axioms: an extension and update of service-dominant logic. J. Acad. Mark. Sci. **44**(4), 5–23 (2016)

Vargo, S.L., Lusch, R.F.: Service-dominant logic 2025. Int. J. Res. Mark. **34**(1), 46–67 (2017)

Vargo, S.L., Maglio, P., Akaka, M.A.: On value and value co-creation: a service systems and service logic perspective. Eur. Manag. J. **26**(3), 145–152 (2008)

Vargo, S.L., Wieland, H., Akaka, M.A.: Innovation through institutionalization: a service ecosystems perspective. Ind. Mark. Manage. **44**, 63–72 (2015)

Vargo, S.L., et al.: A systems perspective on markets—toward a research agenda. J. Bus. Res. **79**, 260–268 (2017)

Windrum, P., García-Goñi, M.: A neo-Schumpeterian model of health services innovation. Res. Policy **37**(4), 649–672 (2008)

Chapter 6
The Innovation Effects of Engaged Actors in Open Platform Ecosystems

María J. Quero and Marco Luzi

Abstract By extending beyond traditional company boundaries, open platform ecosystems introduce novel features worthy of attention, such as the democratic and less predictable nature of innovation involving various actors. Innovation in these ecosystems emerges from interactions between individuals and the system, yet the process that enables this interaction to be innovative requires further understanding. This chapter aims to link actor engagement with resource density in the ecosystem using a Service-Dominant logic framework, focusing on rewards-based crowdfunding due to its highly interactive and interdependent nature. The study develops a model of resource density in open platform ecosystems, illustrating novel connections between resources and innovation mediated by actor engagement. The resulting strategy matrix offers companies actionable innovation strategies for varying actor engagement and resource density levels.

Keywords Innovation ecosystem · Engagement · Actor disposition · Resource density · Crowdfunding

Introduction

The increasingly open and digitized economies have disrupted every dimension of business models. Open platforms emerge as a revolution, connecting actors with disruptive structures that "democratize" the innovation process, making it more complex and, at the same time, more capable to self-adjust in coherence with the

M. J. Quero (✉) · M. Luzi
University of Málaga, Málaga, Spain
e-mail: mqg@uma.es

M. Luzi
e-mail: luzi@uma.es

M. Luzi
University of Agder, Agder, Norway

D. Sarno et al. (eds.), *Complexity and Emergence in Market Ecosystems*, New Economic Windows, https://doi.org/10.1007/978-3-031-81942-1_6

new dynamic concept of market (Mele et al. 2023, 2015; Pels et al. 2023; Storbacka 2019) in which actors have an increasingly active role (Koskela-Huotari and Siltaloppi 2020; Wieland et al. 2016). The present work immerses in the open platform ecosystem context to shed light on how innovation emerges, breaking with concepts like "serendipity" (traditionally used to explain innovation effects) to put in its side a logic (Service-Dominant logic, SDL) and an analysis of the innovation network effects to develop the needed theory to better understand complexity in connection with the emergence in open new platform contexts (Rangaswamy et al. 2022; Nambisan et al. 2019; Gummesson et al. 2018).

Recent literature highlights the need for research on innovation effects in platform business models in connection with a revolution in how innovation emerges in these complex contexts where actors are responsible for leveraging network effects (Srinivasan 2021, p. 5) and calls for theorizing these complex contexts.

The Service-Dominant logic (SDL) offers a theoretical perspective on the process that connects "empowered actors" in open contexts (Wieland et al. 2016; Koskela-Huotari and Siltaloppi 2020; Vargo et al. 2023a) with their capability to trigger innovation effects in the network and make possible emergence and phase transition (Polese et al. 2021).

The contribution provides a theoretical "Open platform ecosystem resource-density management model" that represents a contribution to theory (Chandler et al. 2019; Nambisan et al. 2019) and practice, providing a theoretical structure to understand open platforms dynamics better and improve the capability to design strategies according to the new context, where engaged actors exchange resources in a substantially different way (Mele and Corte 2013; Peters et al. 2014; Lusch and Nambisan 2015a, b).

The context of the research is rewards-based crowdfunding, a type of network that has the characteristics to be considered an open platform ecosystem (Quero & Ventura, 2015, 2019; Quero et al., 2017) and its unique characteristics to connect actors and capabilities have increasingly attracted the attention of researchers and practitioners (Shneor et al., 2024; Bahh-Peprah, 2024; Marciel and Weinberger, 2024; Shneor et al., 2023; Quero & Mele, 2023; Butticé & Ughetto, 2021; Gamble et al., 2019; Fehrer & Nenonen, 2020)

The structure of this chapter is organized as follows. First, we go through the literature review with the remarkable development of three concepts that will be the pillars of the theoretical model (open platform ecosystems, engaged actors connecting resources, and innovation effects). We will develop the Crowdfunding ecosystem resource-density management model in the second stage. Then, we will provide the empirical exemplification in the context of arts reward crowdfunding ecosystems. We will finish with theoretical and empirical contributions and future lines of research.

Literature Review

Resources in Open Platform Ecosystems

Increasing resource density in open platform business models constitutes a challenge for markets (Shi and Shi 2022; Fehrer et al. 2020), as it represents a radical change of strategy in an increasingly collaborative context (Quero et al. 2024). In new forms of business models, the value of new processes is based on liquefying and expanding the scope of resources (Nenonen et al. 2019; Amit and Han 2017; Lusch et al. 2010).

According to Polese et al. (2021), the application of ecosystems to business was fathered by Alderson (1965). The biological metaphor was particularly apt to describe service ecosystems and the idea that all actors contribute to the well-being (value) of the network. Vargo and Lusch (2016, pp. 10–11) describe service ecosystems as "a relatively self-contained, self-adjusting system of resource-integrating actors connected by shared institutional arrangements and mutual value creation through service exchange." This perspective works well within the idea that value is cocreated by all actors, not uniquely by the firm, at the core of Service-Dominant logic (Vargo et al. 2023a; Vargo and Lusch, 2017). Recent literature focuses on the need for affording a complex ecosystem approach to innovation and refers to innovation ecosystems as those that present the requirements to make innovation emerge, especially in digitalized contexts (Poeppelbus et al. 2024; Klimas and Czakon 2022; Grandstand and Holgersson 2020).

This shift away from the centrality of firms and consequent democratization of value creation led to a keen interest in those ecosystems that, most of all, cancel hierarchies among actors: open platform businesses (Fehrer et al. 2018). Open platform ecosystems pierce traditional company boundaries, allowing the potential for more value cocreation from other, typically excluded actors (e.g., Kortmann and Piller 2016; Chesbrough 2007). The *openness* can exist on three levels: (i) the users layer, allowing many actors to be connected; (ii) the technology layer, the technical infrastructure of the platform; and (iii) the provider layer, the coexistence of different classes of actors (Eisenmann et al. 2009; Ondrus et al. 2015).

Windasari and Lin (2021) adopt the SDL perspective to highlight the role of resources in open contexts, and Fehrer et al. (2018) introduce a new business model logic based on open networks and platforms. This logic guides managers to evaluate their platform potential and understand value processes beyond organizational borders. In this perspective, resources in ecosystems emerge as a critical factor, as they represent the network's potential for value creation (Peters et al. 2014; Koskela-Huotari et al. 2016; Peters 2016).

From the SDL ecosystem perspective, two key concepts emerge: integration and interaction (Löbler 2013; Peters et al. 2014). Peters (2016, pp. 3003) makes this distinction by considering two different ways of integrating resources: *homopathic*, with a simple summative or aggregative effect, in which "the resultant effects are identical to the sum of the effects of each of the base resources acting in isolation" and *heteropathic*, when "the presence of emergent properties are new and novel (i.e., new practices) and unpredictable." We will build on these concepts to improve resource density and engagement in open platforms. Fehrer et al. (2018) make a theoretical and empirical contribution to designing strategies to enhance market densities, considering this topic a new research area.

Engaged Actors Connecting Resources

Engagement first appeared in management literature from human resources and workforce motivation: it is the idea of being fully there; as Kahn (1992) puts it, some state of attention that needs to be focused on a specific topic (Schaufeli and Bakker 2004). As Chandler and Lusch (2015) explain, service research borrowed the term but focused on the angle of the interactions between people and their experiences. The first instance of this term is Customer Engagement Behavior (CEB), the cognitive and emotional response of a customer engaging with a firm (Brodie et al. 2019, 2011). Some of the consequences of CEB include the desire to cooperate with the firm, as well as positive word-of-mouth (Verleye et al. 2014) and, in general, including those manifestations of "voluntary resource contributions" (Jaakkola and Alexander 2014, p. 248) and of doing more than just engaging in the transaction. Engagement is still linked with the cognitive and affective state (Schaufeli and Bakker 2004), but it comes from past interactions and influences future interactions. SDL theory (Vargo and Lusch 2017) posits in its fundamental premise number nine (FP9) that "all social and economic actors are resource integrators," and FP6 refers to how resources become valuable for the actors involved in the transaction by setting that "value is co-created by multiple actors." Regarding this process of value cocreation, Storbacka et al. 2016 (pp. 3012) propose actor engagement as a microfoundation for value cocreation and propose the concept of "actor disposition" to bridge the process from theory to practice. In line with this perspective, Brodie et al. (2019, p. 174) define AE as "a dynamic and iterative process that reflects actor dispositions to invest resources in their interactions with other connected actors in a service ecosystem". Adopting this perspective, we consider that engagement manifests through actor engagement behaviors, whereby market actors influence each other's dispositions and behaviors (Fehrer et al. 2020; Alexander et al. 2012) and aim at exploring how to increase innovation emergence (on its different kinds and degrees) through resource density relying on actor's interactions as a strategic step to liquefy and expand the scope of resources (Nenonen et al. 2019; Amit and Han 2017; Lusch et al. 2010).

The shift towards a less customer-centric viewpoint triggered by Service-Dominant logic (Vargo and Lusch 2017) was accompanied by an extension of the concept of engagement beyond the customer and to all actors in the ecosystem (Gummesson 2008; Hillebrand et al. 2015; Brodie et al. 2019). Some papers have supported linking engagement with multi-actor settings (e.g., Li et al. 2017) beyond the traditional, dyadic link between the customer and the firm. Kleinaltenkamp et al. (2019) claim that individual engagement does not suffice and introduce collective engagement as the "multiple actors' shared cognitive, emotional, and behavioral

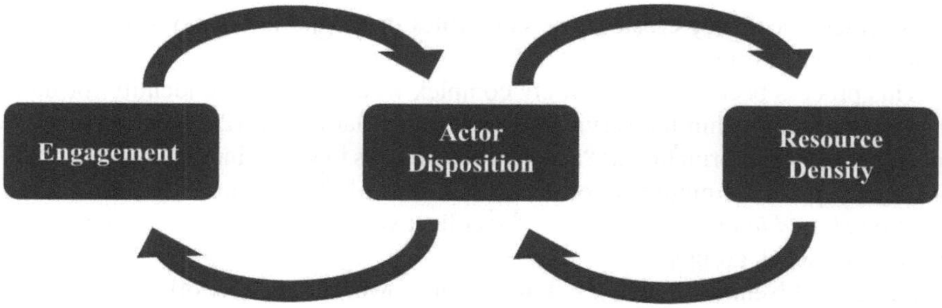

Fig. 6.1 Engagement influences actor disposition and resource integration in open platforms. *Source* The authors

dispositions, as manifested in their interactive efforts devoted to a focal object" (p. 12).

In complex contexts like open platforms, engagement not only positively affects the system's capability to create value but can also be considered strategic support to manage resources and improve the system's value. Actor engagement, which Storbacka et al. (2016, p. 3009) define as "both the disposition of actors to engage and the activity of engaging in an interactive process of resource integration within the institutional context provided by a service ecosystem," will determine the system's capability to grow, maintain, or decrease. A synthesis is reported in Fig. 6.1.

Innovation Effects and Emergence in Ecosystems

Vargo et al. (2023a, p. 3) adopt the systemic perspective of the Service-Dominant logic (SDL) to define emergence as "a phenomenon that arises from the relationships among existing system elements, but that is qualitatively different from and irreducible to them." From this perspective, understanding emergence is a cornerstone of innovation management in open, multisided platforms. This dynamic and collaborative perspective of innovation is in line with the also plastic perspective of markets (Nenonen et al. 2014; Mele et al. 2015, 2023a; Pels et al. 2023), in which it is the responsibility of the engaged actors and their capability to connect resources who will be responsible for triggering innovation (i.e. value cocreation practices) in different forms (Skålen and Gummerus 2023).

As noted in Polese et al. (2021), emergence can contribute to different properties to ecosystems: resources, value, practices, and institutional arrangements -even more, emergence *is* the rise of such new properties in the ecosystem. Meynhardt et al. (2016) draw from synergetics to analyze ecosystem value cocreation as the cyclical action of bottom-up emergence and top-down enslavement. According to this theory, a continuous interaction happens between an ecosystem's micro-level -actors- and the macro-level -social rules-. Actors keep engaging with one another, reshaping the rules (emergence) they created; the same rules direct the interactions and provide coherence (enslavement).

This process becomes increasingly complex when a collective identity (or more than one) forms within the service ecosystem. Taillard et al. (2016) conceptualize service ecosystem formation as "an emergent process in which individual and collective agency [...] are mutually constitutive entities of that system." (p. 2972). They mention *shared intentions* as a proxy for collective agency, i.e., when collections of individuals act as groups.

Fehrer and Nenonen (2020), talking about crowdfunding networks, suggest that creators relinquish control to some extent to build collective agency. Platform ecosystems are seen from a market-oriented perspective, not only as their components but also as markets where users (actors) interact (Eisenmann et al. 2009; Schweiger et al. 2016). The ecosystem grows around the central platform, aggregating different actors.

The Crowdfunding Ecosystem Resource-Density Management Model

Although there is an increasing number of contributions in recent literature highlighting the need for research on actor engagement, resource integration, and emergence in new contexts like open platform ecosystems (Brodie et al. 2019; Fehrer et al. 2018), there is a lack of theoretical and empirical research connecting the three concepts. The present work aims to build on the call for research on how innovation emerges in ecosystems (Polese et al. 2021; Vink et al. 2021; Chandler et al. 2019).

We put our focus on a specific kind of open platform ecosystem: rewards-based crowdfunding.

Rewards-based crowdfunding is a funding method where supporters offer a financial contribution, obtaining some non-monetary reward in return (Mollick 2014). The market size of rewards-based crowdfunding reached 1.25 billion USD in 2020 (Ziegler et al. 2021). Hundreds of crowdfunding platforms exist globally (Shneor et al. 2023a, b), but the most known and studied one in rewards-based crowdfunding is Kickstarter (Shneor and Vik 2020).

The reason for choosing this context comes from both the high revolution that has registered in the last 20 years and their interest in adopting new business models to adapt to new contexts and new forms of demands (Skålen and Gummerus 2023; Edvardsson et al. 2014; Shneor et al. 2024, 2023a, b; Bahh-Peprah , 2024; Butticé and Ughetto 2021; Quero and Mele 2023). Regarding the call for research in this form of open platform, Fehrer and Nenonen (2020, p.449) posit that "research investigating the characteristics of these open and highly dynamic network structures is in its infancy and capabilities critical to operating in them are under-explored." Quero et al. (2017) and Quero and Ventura (2015, 2019) find support in arts crowdfunding contexts to study different forms of value cocreation and see how they account for

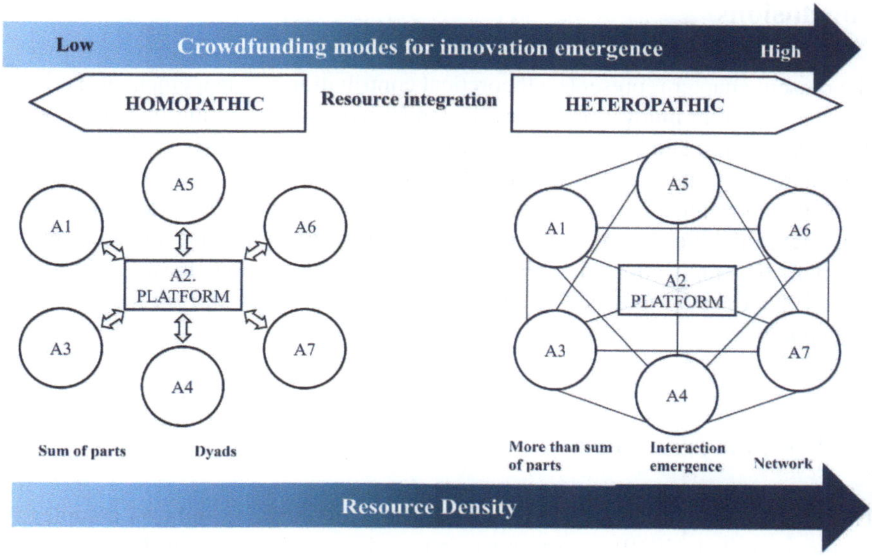

Fig. 6.2 Crowdfunding ecosystem strategies for innovation emergence. *Source* The authors

the conditions to be considered an ecosystem. Edvardsson et al. (2014) highlight the role of resources in the ecosystem, and Skålen and Gummerus (2023) build on the capability of the ecosystem and actor's confluence to give way to innovation (i.e., new practices) in the contexts of the music market.

Figure 6.2 represents the connection between engagement and resources through actor disposition, adopting the SDL perspective of resources as an active asset that will change its capability or potential to bring out innovation depending on the disposition of the actors to manage resources and the capability of the system to make possible heterogeneous actor to converge on a same place/space, making the resources to be exchanged with multiplicative or summative results (Peters 2016).

Crowdfunding literature (Quero and Mele 2023; Quero and Ventura 2015; Belleflamme et al. 2012) identify seven kinds of actors in the crowdfunding context: creators, or the creative core (A1; Ordiani et al. 2011; Belleflamme et al. 2012; Burtch et al. 2013; Quero and Ventura 2015, 2019); platforms (A2; Ordanini et al. 2011; Quero and Ventura 2015, 2019), financing customers (A3; Ordanini et al. 2011; Belleflamme et al. 2012; Burtch et al. 2013; Quero and Ventura 2015, 2019); non-financing customers (A4; Quero and Ventura 2015, 2019); investors (A5; Belleflamme et al. 2012); experts (A6; Belleflamme et al. 2012; Burtch et al. 2013; Quero and Ventura 2015, 2019); and crowdfunding associations (A7).

The crowdfunding context is characterized by the capability to eliminate barriers and make possible actor interactions. The high number of actors involved makes a strategic key issue to evaluate for each: the resources they bring to the system and their engagement with the projects and the platform.

Conclusions

The present chapter represents a theoretical contribution to engagement as a critical factor in reaching innovation and evolution in ecosystems relying on actors' work (Koskela-Huotari and Siltaloppi 2020; Koskela-Huotari et al. 2016). Although there is an increasing interest in shedding light on the way new platforms evolve and self-adjust to new contexts (in coherence with Vargo and Lusch's (2016) definition of an ecosystem), the theoretical and empirical contributions to find specific connections to explain how innovation takes place (in a constant process to evolve and self-adjust) is needed to develop specific strategies that make possible to improve the ecosystem management.

Implications for Theory

This chapter aims to contribute to the challenge of shedding light on complexity. Gummesson et al. (2019) were pioneers in focusing on this issue. The Crowdfunding ecosystem resource-density management model represents a theoretical contribution that gives a perspective to answer the complex questions related to how and why innovation takes place in open platforms (Chandler 2019) and potentially can evolve into Emergence or Phase Transition (Polese et al. 2021). We put our focus on the active role of engaged actors (Koskela-Huotari et al. 2016; Koskela-Huotari and Siltaloppi 2020) and their capability, responsibility, and agency to manage resources and improve resource density, triggering new practices that benefit the system on a recursive engine process. Of course, such a process could also result in negative recursive connections (Mele et al. 2018). As long as the actors present a lower level of engagement, their capability and interest in exchanging resources and resource density will decrease, and the system will put its capability for innovation at a minimum.

The theoretical framework connects three strategic concepts that, linked in the same strategic perspective, represent valuable contributions to theory and practice and open new venues for future research:

According to Koskela-Huotari and Siltaloppi (2020), the actor still needs more attention, especially in terms of their capability to manage resources, make the system evolve, and self-adjust (Mele et al. 2018).

Actor engagement (in connection with actor disposition) is the core of the strategies, according to their capability to manage and potentially trigger new practices in an open multi-actor context.

Resources: the active role of resources, liquefied in the digital business models and co-managed and cocreated within the open network.

Innovation: As initially proposed by Lusch and Nambisan (2015a, b), Service innovation is a collaborative process driven by resource liquefaction, density, and integration. It results in value cocreation between service offer(ers), service beneficiaries, and the rest of the ecosystem's actors.

The positive and negative loops that connect the engaged actor with actor disposition and system resources behave as an "engine," capable of growing (or decreasing) with multiplicative results (Michel et al. 2015). A Service-Dominant logic perspective can better understand and analyze discontinuous innovations, revealing potential breakthrough advances in resource use and customer roles.

Implications for Practice

The research is contextualized in a specific open platform: rewards-based crowdfunding. Literature is increasingly recognizing the need for both: develop more theories that give support to understanding why and how these kinds of business modes are registering such fast growth, in connection with the open economy and new technologies have made it possible to exchange resources among actors in a more accessible and increasingly way, that have brought new ways of creating value into what is called the Sharing Economy (Nambisan et al. 2019; Brem et al. 2019).

The theoretical model can help managers with practical content regarding strategy design based on engagement and resource density in the systems. Innovation Service innovation, as a resource constellation in a service system, is shaped by customers' value cocreation and the interdependencies between resources and schemas (Windasari and Lin 2021; Edvardsson and Tronvoll 2013; Mele et al. 2014).

Figure 6.3 builds on the theoretical model designed for reward crowdfunding open multisided platforms, presenting four possible strategies related to managing the ecosystem towards emergence according to two strategic factors: resource density and actor engagement.

The theoretical matrix identifies four possible positions regarding the management of a crowdfunding ecosystem. Four possible modalities arise regarding actor engagement (low/high) and resource density (low/high). Four possible scenarios can be identified:

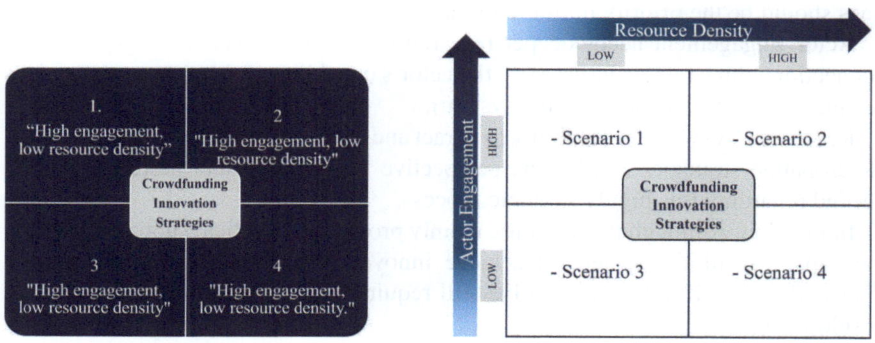

Fig. 6.3 The resource-engagement innovation strategy matrix. *Source* The authors

Scenario 1. *Increase resource density*: the system accounts for the great value of engaged actors committed to the platform and its projects. In this case, it is essential to increase resources to attend to the increasing demand of the highly engaged actors—possible innovation practices emerging. Using the engaged actors to improve resource quality and quantity would be the best way to make the system evolve through new practices arising from actor interaction.

Scenario 2. *Maintain innovativeness*: accounts for the best position for innovation practices and possible Emergence and Phase Transition.

Scenario 3. *Pivot core business*: The low level of resources, in connection with the low engagement of the actors, will probably drive the system's decadence. Strategies to improve actor engagement and resource density would be the strategy to evolve.

Scenario 4. *Engaged actors*: in this position, we find exciting and well-structured contexts in which the richness of the resources offered in the system is not competitive because the actors are probably adopting a very passive attitude toward resources. Digitalization requires systems design that maximizes actor's confluence and participation. The strategy should rely on developing engagement strategies for all the actors to reach the highest position regarding value cocreation in the system.

Limitations and Future Research

Complex contexts like open platforms are still in their infancy regarding theory and practice (Fehrer and Nenonen 2020; Fehrer et al. 2018). However, recent literature has made a great effort to improve the understanding of these complex networks, which are registering unprecedented growth and innovation processes in an increasingly open economy.

In this chapter, we have made a theoretical contribution bridging four strategic concepts: actors, engagement, resource density, and innovation (Fig. 6.4).

The complexity of the open platforms context from this theoretical approach is just one step in a long way, and we have found more questions than answers. Three lines should be the priority in future research:

Actor Engagement needs deeper research to establish levels of engagement in connection with strategic issues (i.e., the actor's capability to propose new practices or other forms of innovation to the system).

Resources: A system's capability to attract and exchange resources at a maximum. Heteropathic strategies can be one perspective for research, but more research is needed regarding this highly strategic process.

Innovation: Although the literature mainly provides theoretical support regarding the capability of the system to achieve innovation, innovation ecosystems and their ability to self-adjust and evolve still require higher theoretical and empirical development.

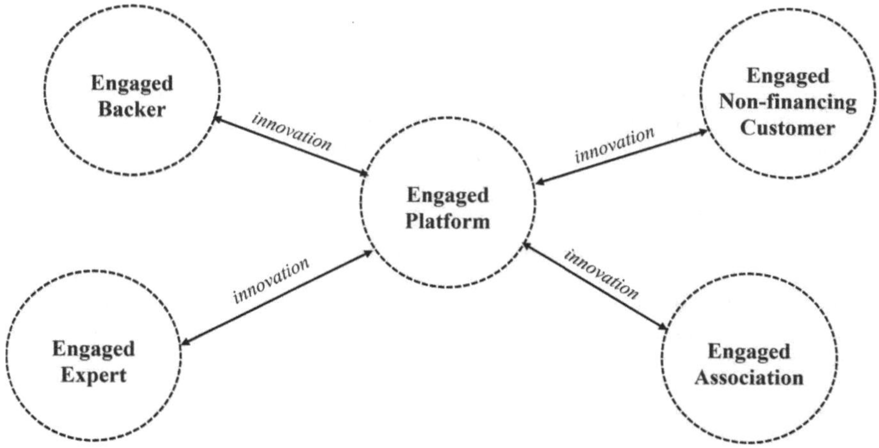

Fig. 6.4 Rewards-based Crowdfunding Platform ecosystem. *Source* The authors

The research context, crowdfunding ecosystems, is also a fast-growing business model that demands the development of theoretical models that improve the capability to design strategies based on theory. The Service-Dominant logic approach accounts for the required tools to explain and afford the system's complexity.

References

Alderson, W.: Dynamic marketing behavior: A functionalist theory of marketing. Richard D. Irwin, Homewood, IL (1965)

Alexander, J., Chase, J., Newman, N., Porter, A., Roessner, J.D.: Emergence as a conceptual framework for understanding scientific and technological progress. In: 2012 Proceedings of PICMET'12: Technology Management for Emerging Technologies. IEEE, pp. 1286–1292. (2012)

Amit, R., Han, X.: Value creation through novel resource configurations in a digitally enabled world. Strateg. Entrep. J. **11**(3), 228–242 (2017)

Baah-Peprah, P., Shneor, R., Munim, Z.H.: In this together: The antecedents and implications of crowdfunding community identification and trust. Ventur. Cap. Ahead-of-print (2024)

Baker, T., Nenonen, S.: Markets changing, changing markets: institutional work as market shaping. Ind. Mark. Manage. **89**, 13–20 (2020). https://doi.org/10.1016/j.indmarman.2020.02.004

Belleflamme, P., Lambert, T., Schwienbacher, A.: Individual crowdfunding practices. http://creationetfinancement.files.wordpress.com/2012/12/individual-crowdfunding-practices.pdf(2012). Last accessed (2022)

Brem, A., Bilgram, V., Marchuk, A.: How crowdfunding platforms change the nature of user innovation—from problem-solving to entrepreneurship. Technol. Forecast. Soc. Chang. (2019). https://doi.org/10.1016/J.TECHFORE.2017.11.020

Brodie, R.J., Hollebeek, L.D., Jurić, B., Ilić, A.: Customer engagement: conceptual domain, fundamental propositions, and implications for research. J. Serv. Res. **14**(3), 252–271 (2011)

Brodie, R.J., Fehrer, J.A., Jaakkola, E., Conduit, J.: Actor engagement in networks: defining the conceptual domain. J. Serv. Res. **22**(2), 173–188 (2019)

Burtch, G., Ghose, A., Wattal, S.: An empirical examination of the antecedents and consequences of contribution patterns in crowd-funded markets. Inf. Syst. Res. **24**(3), 499–519 (2013)

Butticé, V., Ughetto, E.: What, where, who and how? a bibliometric study of crowdfunding research. IEEE Trans. Eng. Manage. **70**(9), 3078–3099 (2021)

Chandler, J.D., Lusch, R.F.: Service systems: A broadened framework and research agenda on value propositions, engagement, and service experience. J. Serv. Res. **18**(1), 6–22 (2015)

Chandler, J.D., Danatzis, I., Wernicke, C., Akaka, M.A., Reynolds, D.: How does innovation emerge in a service ecosystem? J. Serv. Res. **22**(1), 75–89 (2019). https://doi.org/10.1177/109467051 8797479

Chesbrough, H.W.: Why companies should have open business models. MIT Sloan Manag Rev (2007)

Edvardsson, B., Tronvoll, B.: A new conceptualization of service innovation grounded in S-D logic and service systems. Int. J. Qual. Serv. Sci. **5**(1), 19–31 (2013). https://doi.org/10.1108/175666 91311316220

Edvardsson, B., Kleinaltenkamp, M., Tronvoll, B., McHugh, P., Windahl, C.: Institutional logics matter when coordinating resource integration. Mark. Theory **14**(3), 291–309 (2014). https:// doi.org/10.1177/1470593114534343

Eisenmann, T.R., Parker, G., Van Alstyne, M.: Opening platforms: how, when, and why. In Platf. Mark. Innov. pp. 131–162 (2009)

Fehrer, J.A., Woratschek, H., Brodie, R.J.: A systemic logic for platform business models. J. Serv. Manag. **29**(4), 546–568 (2018)

Fehrer, J.A., Nenonen, S.: Crowdfunding networks: Structure, dynamics, and critical capabilities. Ind. Mark. Manage. **88**, 449–464 (2020)

Gamble, J.R.: Marketing madness or financial folly? Implementing equity crowdfunding in the recording industry. Eur. J. Mark. **53**(3), 412–441 (2019)

Gummesson, E., Mele, C., Polese, F.: Complexity and viability in service ecosystems. Mark. Theory **18**(2), 187–207 (2018). https://doi.org/10.1177/1470593118774201

Granstrand, O., Holgersson, M.: Innovation ecosystems: a conceptual review and a new definition. Technovation **90–91**, 1–12 (2020)

Gummesson, E., Mele, C., Polese, F.: Complexity and viability in service ecosystems. Mark. Theory **18**(2), 187–207 (2019). https://doi.org/10.1177/1470593118774201

Gummesson, E.: Quality, service-dominant logic, and many-to-many marketing. The TQM Journal **20**(2), 143–153 (2008). https://doi.org/10.1108/17542730810857372

Hillebrand, B., Driessen, P.H., Koll, O.: Stakeholder marketing: theoretical foundations and required capabilities. J. Acad. Mark. Sci. **43**(4), 411–428 (2015). https://doi.org/10.1007/s11747-015-0424-y

Jaakkola, E., Alexander, M.: The role of customer engagement behavior in value cocreation: a service system perspective. J. Serv. Res. **17**(3), 247–261 (2014)

Kahn, W.A.: To be fully there: psychological presence at work. Hum. Relat. **45**(4), 321–349 (1992)

Kleinaltenkamp, M., Karpen, I.O., Plewa, C., Jaakkola, E., Conduit, J.: Collective engagement in organizational settings. Ind. Mark. Manage. **80**, 11–23 (2019)

Klimas, P., Czakon, W.: Species in the wild: a typology of innovation ecosystems. Eur. J. Innov. Manag. **25**(2), 267–289 (2022). https://doi.org/10.1108/EJIM-02-2021-0072

Kortmann, S., Piller, F.: Open business models and closed-loop value chains: Redefining the firm-consumer relationship. Calif. Manage. Rev. **58**(3), 88–108 (2016)

Koskela-Huotari, K., Edvardsson, B., Jonas, J., Sörhammar, D., Witell, L.: Innovation in service ecosystems—breaking, making, and maintaining institutionalized rules of resource integration. J. Bus. Res. **69**, 2964–2971 (2016). https://doi.org/10.1016/J.JBUSRES.2016.02.029

Koskela-Huotari, K., Siltaloppi, J.: Rethinking the actor in service research: toward a processual view of identity dynamics. J. Serv. Theory Pract. **30**(4/5), 437–457 (2020). https://doi.org/10.1108/JSTP-11-2018-0245

Li, L.P., Juric, B., Brodie, R.J.: Dynamic multi-actor engagement in networks: the case of United Breaks Guitars. J. Serv. Theory Pract. **27**(4), 738–760 (2017)

Löbler, H.: Service-dominant networks: an evolution from a service-dominant logic perspective. J. Serv. Manag. **24**(4), 255–282 (2013)

Lusch, R.F., Vargo, S.L., Tanniru, M.: Service, value networks, and learning. J. Acad. Mark. Sci. **38**(1), 19–31 (2010)

Lusch, R.F., Nambisan, S.: Service innovation: a service-dominant logic perspective. MIS Quarterly **39**(1), 155–175 (2015). https://doi.org/10.25300/misq/2015/39.1.07

Lusch, R., Nambisan, S.: Service innovation: a service-dominant logic perspective. MIS Q. **39**, 155–175 (2015). https://doi.org/10.25300/misq/2015/39.1.07

Marciel, A.F., Weinberger M.F.: Crowdfunding as a market-fostering gift system. J. Consum. Res. 1221, 1242 (2024). https://doi.org/10.1093/jcr/ucad052

Mele, C., Colurcio, M., Russo-Spena, T.: Research traditions of innovation: goods-dominant logic, the resource-based approach, and service-dominant logic. Manag. Serv. Qual. **24**(6), 612–642 (2014). https://doi.org/10.1108/MSQ-10-2013-0223

Mele, C., Nenonen, S., Pels, J., Storbacka, K., Nariswari, A., Kaartemo, V.: Shaping service ecosystems: exploring the dark side of agency. J. Serv. Manag. **29**(4), 521–545 (2018). https://doi.org/10.1108/JOSM-02-2017-0026

Mele, C., Corte, V.: Resource-based view and service-dominant logic: similarities, differences and further research. J. Bus. Mark. Manag. **6**, 192–213 (2013)

Mele, C., Pels, J., Storbacka, K.: A holistic market conceptualization. J. Acad. Mark. Sci. **43**, 100–114 (2015). https://doi.org/10.1007/s11747-014-0383-8

Mele, C., Pels, J., Spano, M., Di Bernardo, I.: Emergent understandings of the market. Ital. J. Mark. 1–25 (2023). https://doi.org/10.1007/s43039-022-00062-z

Meynhardt, T., Chandler, J.D., Strathoff, P.: Systemic principles of value cocreation: synergetics of value and service ecosystems. J. Bus. Res. **69**(8), 2981–2989 (2016)

Mollick, E.: The dynamics of crowdfunding: an exploratory study. J. Bus. Ventur. **29**(1), 1–16 (2014)

Nambisan, S., Siegel, D., Kenney, M.: On open innovation, platforms, and entrepreneurship. Strateg. Entrep. J. (2019). https://doi.org/10.1002/sej.1300

Nenonen, S., Kjellberg, H., Pels, J., Cheung, L., Lindeman, S., Mele, C., Santos, L., Storbacka, K.: A new perspective on market dynamics: market plasticity and the stability-fluidity dialectics. Mark. Theory **14**(3), 269–289 (2014). https://doi.org/10.1177/1470593114534342

Nenonen, S., Storbacka, K., Windahl, C.: Capabilities for market shaping: triggering and facilitating increased value creation. J. Acad. Mark. Sci. **47**, 617–639 (2019). https://doi.org/10.1007/s11747-019-00643-z

Ondrus, J., Gannamaneni, A., Lyytinen, K.: The impact of openness on the market potential of multisided platforms: a case study of mobile payment platforms. J. Inf. Technol. **30**(3), 260–275 (2015)

Ordanini, A., Miceli, L., Pizzetti, M., Parasuraman, A.: Crowd-funding: transforming customers into investors through innovative service platforms. J. Serv. Manag. **22**(4), 443–470 (2011)

Pels, J., Mele, C., Spano, M.: From market driving to market shaping: impact of language shift. J. Bus. Ind. Mark. **38**(1), 155–169 (2023). https://doi.org/10.1108/JBIM-10-2021-0503

Peters, L.D., Pressey, A.D., Vanharanta, M., Johnston, W.J.: Theoretical developments in service-dominant logic: propositions and directions for future research. J. Bus. Res. **67**(8), 1341–1352 (2014)

Peters, L.D.: Heteropathic versus homopathic resource integration and value cocreation in service ecosystems. J. Bus. Res. **69**(8), 2999–3007 (2016). https://doi.org/10.1016/j.jbusres.2016.02.033

Poeppelbuss, J., Ebel, M., Anke, J.: Iterative uncertainty reduction in multi-actor smart service innovation. Electron. Mark. **32**(3), 599–627 (2024). https://doi.org/10.1007/s12525-021-00500-4

Polese, F., Payne, A., Frow, P., Sarno, D., Nenonen, S.: Emergence and phase transitions in service ecosystems. J. Bus. Res. **127**, 25–34 (2021)

Quero, M.J., Mele, C.: Balanced centricity: A joint institutional logic within open, collaborative ecosystems. J. Bus. Ind. Mark. **38**(2), 384–394 (2023). https://doi.org/10.1108/JBIM-12-2021-0572

Quero, M.J., Ventura, R.: The role of balanced centricity in the Spanish creative industries adopting a crowd-funding organizational model. J. Serv. Theory Pract. **25**(2), 122–139 (2015). https://doi.org/10.1108/JSTP-09-2013-0182

Quero, M.J., Ventura, R., Kelleher, C.: Value-in-context in crowdfunding ecosystems: how context frames value cocreation. Serv. Bus. **11**, 405–425 (2017)

Quero, M.J., Ventura, R.: Value proposition as a framework for value creation in crowdfunding ecosystems. Mark. Theory **19**(1), 47–63 (2019)

Quero, M.J., Ruiz-Alba, J.L., Mele, C., Tronvoll, B., Ventura, R.: Internationalization of innovation practices: actors and linkages in crowdfunding ecosystems. Eur. J. Int. Manag. (2024). https://doi.org/10.1504/EJIM.2023.10063669

Rangaswamy, A., Moch, N., Felten, C., van Bruggen, G.H., Wieringa, J., Wirtz, J.: The role of marketing in digital business platforms. J. Interact. Mark. **51**, 72–90 (2022). https://doi.org/10.1016/j.intmar.2020.04.006

Schaufeli, W.B., Bakker, A.B.: Job demands, job resources, and their relationship with burnout and engagement: a multi-sample study. J. Organ. Behav: Int. J. Ind. Occup. Organ. Psychol. Behav. **25**(3), 293–315 (2004)

Schweiger, A., Nagel, J., Böhm, M., Krcmar, H.: Platform business models. Digit. Mobil. Platf. Ecosyst. 66–90. (2016)

Shi, W., Shi, Y.: Unpacking the process of resource allocation within an entrepreneurial ecosystem. Res. Policy **50**(9), 104378 (2022)

Shneor, R., Vik, A.A.: Crowdfunding success: a systematic literature review 2010–2017. Balt. J. Manag. **15**(2), 149–182 (2020)

Shneor, R., Zhao, L., Goedecke, J.F.M.: On relationship types, their strength and reward are crowdfunding backer behavior. J. Bus. Res. **154**, 113294 (2023)

Shneor, R., Wenzlaff, K., Boyko, K., Baah-peprah, P., Odorovic, A., Okrimenko, O.: The European Crowdfunding Market Report 2023. University of Agder- Crowdfunding Research Center, Kristiansand (2024)

Shneor, R., Wenzlaff, K., Boyko, K., Baah-Peprah, P., Odorović, A., Okhrimenko, O.: The European Crowdfunding Market Report (2023)

Skålen, P., Gummerus, J.: Conceptualizing services and service innovation: a practice theory study of the Swedish music market. J. Serv. Res. **26**(1), 83–102 (2023). https://doi.org/10.1177/10946705211062693

Srinivasan, R.: Platform business models. Springer Singapore (2021). https://doi.org/10.1007/978-981-99-4910-6

Storbacka, K., Brodie, R.J., Böhmann, T., Maglio, P.P., Nenonen, S.: Actor engagement as a microfoundation for value cocreation. J. Bus. Res. **69**(8), 3008–3017 (2016). https://doi.org/10.1016/j.jbusres.2016.02.034

Storbacka, K.: Actor engagement, value creation, and market innovation. Ind. Mark. Manage. **18**, 4–10 (2019)

Taillard, M., Peters, L.D., Pels, J., Mele, C.: The role of shared intentions in the emergence of service ecosystems. J. Bus. Res. **69**(8), 2972–2980 (2016)

Verleye, K., Gemmel, P., Rangarajan, D.: Managing engagement behaviors in a network of customers and stakeholders: evidence from the nursing home sector. J. Serv. Res. **17**(1), 68–84 (2014)

Vargo, S.L., Lusch, R.F.: Institutions and axioms: an extension and update of service-dominant logic. J. Acad. Mark. Sci. **44**, 5–23 (2016)

Vargo, S.L., Lusch, R.F.: Service-dominant logic 2025. Int. J. Res. Mark. **34**(1), (2017)

Vargo, S.L., Peters, L., Kjellberg, H., Koskela-Huotari, K., Nenonen, S., Polese, F., Sarno, D., Vaughan, C.: Emergence in marketing: an institutional and ecosystem framework. J. Acad. Mark. Sci. **51**, 2–22 (2023a). https://doi.org/10.1007/s11747-022-00849-8

Vargo, S.L., Wieland, H., O'Brien, M.: Service-dominant logic as a unifying theoretical framework for the re-institutionalization of the marketing discipline. J. Bus. Res. 164, 113965 (2023b). https://doi.org/10.1016/j.jbusres.2023.113965

Vink, J., Edvardsson, B., Wetter-Edman, K., Tronvoll, B.: Service ecosystem design: Propositions, process model, and future research agenda. J. Serv. Res. **24**(2), 168–186 (2021). https://doi.org/10.1177/1094670520952537

Wieland, H., Koskela-Huotari, K., Vargo, S.L.: Extending actor participation in value creation: an institutional view. J. Strateg. Mark. **24**(3–4), 210–226 (2016). https://doi.org/10.1080/0965254X.2015.1095225

Windasari, N., Lin, F.: Explicating open innovation using service-dominant logic: strategic implications. Int. J. Serv. Sci. Manag. Eng. Technol. **12**, 78–98 (2021). https://doi.org/10.4018/IJSSMET.2021030105

Ziegler, T., Shneor, R., Wenzlaff, K., Suresh, K., Paes, F.F.C., Mammadova, L., Knaup, C.: The 2nd global alternative finance market benchmarking report (2021)

Chapter 7
The Value Depletion Circle in Higher Education and Its Evolution to Nowhere: SDL Rescue

Montserrat Díaz-Méndez, Michael Saren, Moisés Rubén Zamora-Ramos, and Antonio Chamorro-Mera

Abstract This chapter examines the Higher Education (HE) ecosystem and the unintended negative consequences of importing market practices from business in the mistaken belief that they will similarly improve HE efficiency. This is often described as so-called 'marketization' which can result in negative consequences when applied to essential public services such as Healthcare or Education. We argue that an element or actor essential to one ecosystem may not function in the same way in another one, because each ecosystem has unique characteristics leading to different dynamics. Marketization in HE is considered an ecosystem polluting element implying the consideration of students as customers what involves serious consequences for society development that are addressed here. We explain and apply the concept of Value Depletion Circle (VDC) to analyse the detrimental effects of marketization in HE and show how the Service Dominant Logic approach could counteract the VDC or mitigate its negative consequences.

Keywords Higher Education · Service ecosystem · Service Dominant Logic · Students evaluation surveys · Value co-creation

---

M. Díaz-Méndez (✉) · A. Chamorro-Mera
University of Extremadura, Badajoz, Spain
e-mail: mdmendez@unex.es

A. Chamorro-Mera
e-mail: chamorro@unex.es

M. Saren
University of Birmingham, University of Leicester (UK), Leicester, UK
e-mail: majs1@le.ac.uk

M. R. Zamora-Ramos
Autonomous University of the State of Hidalgo, Pachuca, Mexico
e-mail: moises_zamora@uaeh.edu.mx

D. Sarno et al. (eds.), *Complexity and Emergence in Market Ecosystems*, New Economic Windows, https://doi.org/10.1007/978-3-031-81942-1_7

Introduction

In 1946, the Argentine government imported pairs of beavers from Canada to breed them in Tierra del Fuego seeking the same benefits that the species meant for Canada. However, today, there are over 150,000 beavers in the territory, and they are considered an uncontrollable plague. These beavers significantly disrupt the ecosystem, and their presence threatens the overall functioning of the local environment. While beavers are not problematic in Canada, removing them from their native habitat and introducing them into a "similar" ecosystem was a serious mistake. The government is now attempting to eradicate this species although this is extremely costly and complicated to implement; and even if successful, the damage to the ecosystem remains severe (Baldini et al. 2008; Schiavini et al. 2016).

This illustrates that an element or actor essential to one ecosystem may not function in the same way in another, because each ecosystem has unique characteristics, leading to different dynamics. The road to hell is paved with good intentions, say the adage. This chapter deals with the Higher Education (HE) ecosystem and the unintended negative consequences of importing market practices from business in the mistaken belief that they will similarly improve HE efficiency.

The Concept of Service Ecosystem

In biology, an ecosystem is defined as a complex network of interactions among living organisms (such as animals, plants and microorganisms) and their physical environment (including soil, water and air), creating a dynamic balance that supports the survival and evolution of species within their habitat. Taking this concept to the field of service, Vargo and Lusch (2014, p. 24) describe a service ecosystem as a "relatively autonomous and self-adjusting system of resource-integrating agents, connected by shared institutional arrangements and mutual value creation through service exchange". This ecosystem of services, like biological ones, is composed of living and non-living elements, which impact directly or indirectly, the provision of services through an interconnected process. This process includes multiple interdependent relationships and institutional arrangements, which should reflect the balance and evolution observed in natural ecosystems.

Specifically, this chapter explores how the Service Ecosystem approach influences Higher Education (HE) by examining the complex dynamics of the educational ecosystem and the forces that shape it. First, HE is described as a dynamic ecosystem composed of diverse institutions, actors and interconnected practices. It highlights that marketization is a significant challenge facing this ecosystem, which negatively affects both the quality and accessibility of education, distorting and polluting its complex dynamics. The chapter focuses on and further analyses these unintended negative consequences of importing market practices from business in the mistaken belief that they will similarly improve HE efficiency. On the contrary, the authors contend that the wholesale importation of business methods sets in train a *value depletion circle* (Díaz-Méndez & Saren 2019) which will eventually result in a distortion of the role of HE in society as a central pillar of human development (Díaz-Méndez et al. 2017). Service-Dominant Logic (SDL) is employed as a valuable conceptual approach to explore and identify appropriate measures to restore balance in the educational ecosystem. By adopting this service perspective in order to foster value co-creation processes HE institutions can mitigate against value depletion and refocus on the well-being and holistic development of students.

Higher Education Service Ecosystem

HE has long been recognized as a fundamental pillar for the social and economic development of individuals, societies and nations. In an increasingly complex and globalized world, the formation of skilled and critical citizens is essential to meet the current challenges. Hence, HE not only provides specific knowledge and skills but also plays a crucial role in the formation of individuals capable of contributing to the progress and well-being of society (Díaz-Méndez et al. 2019; Gao et al. 2019; Krstić et al. 2020; Sanchez-Carrillo et al. 2021). In this sense, the Service Ecosystem approach has been highly related to HE worldwide (Nie et al. 2019; Polese et al. 2022) to better understand the interconnected elements as a complex network of interactions among actors (Autio and Thomas 2018; Vargo and Lusch 2014), where internal organizational dynamics intertwine with external influences impacting the educational experience (Grönroos 2011; Sklyar et al. 2019).

Applying the Service Ecosystem approach to HE, involves adopting a complex and interdisciplinary perspective that recognizes the intricate and interconnected nature of the system. According to Nie et al. (2019) and Richter et al. (2015), this approach acknowledges that HE is not isolated but rather part of a broader network that includes several distinct actors. These actors and stakeholders encompass students, faculty, government, alumni, prospective students, student families, university councils, NGOs, private institutions, evaluation agencies and society at large (Díaz-Méndez and Gummesson 2012). Within the HE service ecosystem, these entities are deeply interconnected and interdependent in some way or another We are aware that not all HE systems are alike, they differ in structure, relation to the State, organization, governance, etc. but over recent years the tendency to marketize institutions and processes has been a global phenomenon common to HE systems in most countries.

The interactions among these actors (living elements) are influenced by non-living elements (financial resources, legislative frameworks, cultural norms, management practices, and the reputations of both the country and the university itself) that shape and guide their engagements in distinctive ways (Díaz-Méndez et al. 2017; Koskela-Huotari and Vargo 2016). The diversity of these operant resources in the HE ecosystem adds another layer of complexity; these institutions vary greatly in size, mission, focus and structure, some are publicly funded while others are privately operated, ranging from expansive research universities to small colleges (Marginson 2006). Based on the above, the value of educational experience can be seen to be co-created through networks of relationships and interactions among actors, shaped by operand resources for value co-creation.

It is relevant to note that the value co-created in business service ecosystems, often becomes apparent immediately during interaction and upon completion. However, in the HE service ecosystem, this value may emerge years or even decades after students finish their studies. Value may emerge when students apply the knowledge they acquired and the skills they developed during their academic training. In addition, this co-created value does not occur spontaneously or chaotically; rather, it necessitates a degree of coordination. This is where institutions and their arrangements become crucial. Institutions encompass more than formal entities such as governments or organizations; they also include norms, rules, practices and shared meanings adopted by actors, to establish a framework that allows interactions to develop effectively (Quero and Ventura 2022).

In this sense, the Higher Education Institutions (HEIs) are the central institutions for the coordination of the dynamic systems, processes and technologies in the HE service ecosystem (Autio and Thomas 2018; Koskela-Huotari and Vargo 2016). Their primary mission is to foster a comprehensive and supportive environment that promotes academic success by guiding students to achieve their professional and personal goals (Barile et al. 2016; Voropai et al. 2019). These HEIs not only create value through their educational offerings and policies but they are also influenced, directly or indirectly, by their inter-institutional relationships with other public and private entities (Ferrary and Granovetter 2009; Mason and Brown 2014).

These inter-institutional relationships result in formal (laws, rules, agreements and regulations) and informal (customs, traditions and cultures) institutional arrangements that primarily seek to coordinate the service experience. However, a crucial question arises here: do all institutional arrangements effectively coordinate the relations of all actors? Some of these arrangements have incentivized or allowed the integration of generic market practices from other service ecosystems into the HE ecosystem. This has resulted in the "pollution" of the harmonious development of the educational ecosystem, as some practices introduced that have proven benefits in other contexts are not effectively adapted to the educational environment, leading the ecosystem to a state of marketization with worrying long term consequences for societies' development.

Marketization of Higher Education

In recent decades, HE has undergone a significant transformation due to the growing influence of the classic market logic in its structure and operation. This phenomenon, known as the "marketization" of education, has generated several challenges and concerns around its social value and fundamental purpose (Judson and Taylor 2014; Schelble 2006). In an environment of current market competition, HEIs, both public and private, are often forced to adopt generic business practices and marketing strategies to attract and retain students. This trend can lead to a utilitarian perspective of education, by adopting a simple market metaphor in which students are considered as customers, teachers act as service providers and education is regarded as a product to be produced and delivered (Díaz-Méndez and Gummesson 2012; Elsharnouby 2015; Seeman and O'Hara 2006). It is important to clarify that marketization is not marketing per se. The concept of HE marketization refers to the process of transforming HE service ecosystem into a market-oriented system where students are regarded as passive customers to be supplied with knowledge and skills as if they were transferable commodities. Universities are seen as producers and education is seen as a product that can be bought and sold (Judson and Taylor 2014). This process has entailed the introduction of competition, consumer choice and market-oriented decision-making in the provision of academic activities, driven by several factors, including the rising student demand and the diminishing role of public funding (Banwait and Hancock 2021; Schelble 2006). It is important to highlight that marketization is not negative per se. It can result in negative consequences however particularly when applied to services that correspond to the essence of human development and coexistence such as health care, education or law and order. Thus, it is partial and inappropriate to regard patients, students or citizens simply as 'service customers'.

One of the reasons that has led to this marketization is the response to increased competition among universities to attract and keep students. The growing demand for HE and the expansion of universities have created a competitive market for educational services. Universities must compete to attract students and to do so they employ a variety of commercial strategies that seek to capture the largest number of prospects. We are not again these marketing practices designed to attract students. However, many of these HEIs are focusing their efforts on immediate student satisfaction, as an attraction strategy, sometimes to the detriment of educational quality. This can lead them to become degrees factories, where the main objective is to have satisfied students, although not necessarily well-trained academically (Judson and Taylor 2014; Díaz-Méndez et al. 2019; King and Bunce 2020; Schelble 2006).

This shift has led educational institutions to divert their focus away from the primary objective of education: fostering learning and the development of the skills of the students. HEIs' goal has moved from students' learning to students' perceived immediate satisfaction. Within this framework, the utility of HE has come under scrutiny by some sectors of society (employers, families, prospects, etc.), accompanied by a declining interest among young people to make efforts to get a university degree. This trend stems from the recognition that, even with a degree, gaining access to meaningful employment remains challenging (Dollinger et al. 2018; Schelble 2006), perhaps because the main focus is to keep students 'happy' during their educational experience, neglecting the prioritization of meaningful learning and the longer time-dimension involved. This marketization is integrated by different polluting practices brought from other ecosystems, supported by the same institutions and their arrangements, as shown below.

Higher Education Service Ecosystem Pollution

In the field of HE, some detrimental practices imported from markets include the use of customer (aka student) evaluation surveys. Originally intended to gauge student satisfaction, these surveys are often utilized as metrics for educational quality and the performance of departments and individual teachers, and even to assess whether faculty members possess adequate knowledge or training for teaching. In the most extreme cases, these surveys become the basis for determining whether teachers should keep their positions (Díaz-Méndez et al. 2022).

Measuring service quality through customer satisfaction is a desired practice in other ecosystems. However, these surveys are not intended for students to assess the qualifications, knowledge or achievements of faculty members, as students-in their role as learners- are not qualified to make such assessments. Furthermore, these surveys cannot effectively evaluate the long-term effectiveness of teaching methods employed during the educational process, as students often lack the perspective needed to make such judgments even years after graduation (Díaz-Méndez et al. 2017).

In addition, a satisfaction survey does not necessarily correlate with an objective assessment of the educational service. In the short term, a student may express satisfaction despite not having learned, or conversely, may express dissatisfaction even when significant learning has occurred. Just as many medical treatments cause short-term discomfort but aim to improve long-term health, many educational strategies and tools are implemented to develop both generic and specific skills necessary to improve the capabilities of the students, rather than catering to their immediate comfort and satisfaction (Bovill et al. 2011; Bovill 2020; Dollinger et al. 2018).

When evaluating the performance of teachers, relying solely on the immediate satisfaction of students during the educational experience will not provide a comprehensive measure of their overall educational value. Over time, as students come to appreciate the enduring impact of education on their personal and professional development, their perspective on these evaluations often evolves as they mature and gain greater critical ability. Consequently, it is inappropriate to apply the same satisfaction surveys used for assessing restaurant dining experiences, hotel stays or household appliances to the evaluation of educational services. The true value of education often only becomes evident long after university with the former students' personal reflection, application and experience, which highlights the need for more nuanced, holistic and long-term approaches to the assessment of educational quality and effectiveness.

However, even with these limitations, the use of surveys has not slowed down their application; on the contrary, they are widely used, even when the HE service ecosystem is an intrinsically complex field and simplifying this complexity through numerical scoring can be problematic. The standardization required for these surveys often fails to capture the entirety of the educational environment, omitting numerous factors and actors involved in the teaching process (Berbegal-Mirabent and Ribeiro-Soriano 2015). This simplification can lead to erroneous interpretations and have serious and unfair repercussions for teachers in the short term, but also for the students themselves in the long term, affecting crucial aspects of their professional opportunities.

On the other hand, because of this polluting practice, grade inflation and loss of university rigor have resulted (Judson and Taylor 2014; Díaz-Méndez et al. 2022). Instructors may feel that they must give higher grades to receive favorable evaluations, which in turn may affect their chances for promotion and job security. This has led universities to a dynamic to be forced to meet student satisfaction through higher grades with less effort. This creates a bias in teaching evaluations, as teachers may feel pressured to adjust their grading standards to keep students "happy" and thereby improve retention and enrollment rates. The commitment to society is apparently left behind with worrying consequences for future generations. It also damages the institution's image and undermines the reputation of university teachers. The fact that in this "import from commercial markets' quality evaluation process" teachers have now become the subject of students' evaluation sounds senseless and introduces confusion to the role of teachers and the University itself. Traditionally it was teachers who evaluate students with no further consequences for anyone. If a student fails an exam, he or she can take other opportunities to pass it but, if teachers fail students' "evaluation", they face consequences for their salary and other working conditions to the extent, in many universities, to be able to lose their job. This has led HE to a critically polluted state, as it has become a commercial product where quality criteria are based on standards external to academia, rather than focusing on the pursuit of truth and knowledge, because when grades no longer accurately reflect the development of students, the value of a university is undermined.

Polluting practices harm HE service ecosystem. Graduating students with inflated grades may lack the necessary competencies. In addition, they are likely to misperceive their learning experience as satisfactory, what leads to an incorrect assessment of educational quality, creating a false sense of competence and professional success among both students and employers. As a result, employers may find underprepared graduates with high grades but lacking the necessary professional skills, which contributes to a widespread devaluation of university degrees and increasing distrust in the education system. Another polluting factor related to the explained above is the consideration of students' good grades as a "success rate" of teachers. A teacher's success is measured in many universities through their students pass rate, thus making teachers entirely responsible for students' success or failure (Díaz-Méndez et al. 2017). But not only that - this means that teachers are responsible for each of their students' grades regardless of students' own contribution which is not taken into account.

We see other examples of business practices introduced into HE such as those explained by Fleming (2021) regarding the online delivery of pre-recorded classes. In one case he cites students were offered courses taught on-line by teachers they could not contact, which they eventually discovered was due to the fact that those teachers had passed away. Fleming makes insightful reflections about the limits of the current university and states that "there's no doubt that an 'Edu-Factory' stands today where the traditional university once did" and quotes Bill Reading's classic The University of Ruins (1996). In the 1990's Reading highlighted that what he considered to be the central mission of HE, to inculcate the exercise of critical judgment, was in danger of being killed off. And if that happened, the university "no longer participates in its historical project for humanity" (Readings 1996, 5–6). Since then marketization of HE has extended exponentially in all parts of the world and regardless of the political ideology of the government. It appears that this ideology has been unquestionably adopted by managers of HE institutions responsible for designing "the rules of the game".

If students were to recognize their own key role as co-creators of the value of their educational experience, playing an active part in their learning process, they might start to think of themselves as co-responsible for the "service outcome" and thus counteract their expectation to be treated as customers.

This co-creation approach has been addressed in the service literature in the last decades. In this chapter, we focus on the Service-Dominant logic (SDL) approach which considers that co-creation of value can be instrumental in incorporating students as active actors in the creation of educational value. However, the value co-creation perspective is applied to other type of organizations and although from a theoretical viewpoint it may seem easy to adopt, in practice particularities and the complexity of the specific service system may hinder the purpose of co-creation.

The Value Depletion Circle in Higher Education

The irony is that private sector businesses from whom this model of marketization has been appropriated by HE do not themselves regard their relationships with customers as always entirely value creating or positively co-creating. For example, companies are well aware that their commercial operations can fail to satisfy customers on occasions, sometimes through no fault of their own. Thus, they set in place various service recovery processes such as product recall, compensation, refunds, etc. They also recognize that their everyday interaction with customers relies not only on their own creation of value or the collaborative value co-creation with customers, suppliers, etc. but also that it is customers perception of the value that is produced or delivered to them that matters. For service operations this means that value is phenomenologically and individually experienced (Vargo and Lusch 2014) and each customer makes their own often differing assessment of the value. Importantly, a critical component of this value perception is the relationship between service provider and the customer.

One study of value co-creation in the advertising industry investigated how and why clients' perception of low value of advertising agencies services was a general problem for the sector (Díaz-Méndez and Saren 2019). Reasons included inconsistency of outcome of advertising and client expectations, saturation, lack of effectiveness, and lack of tools to measure the results of the advertising service and wrong expectations. This study also found that *complex clients* could actually hinder the value co-creation process. Not all client relationships resulted in positive value creation because some eventually produce detrimental or negative value. The authors conceptualized this process by which value is reduce as the *value depletion circle (VDC)*. Advertising agencies often reported that dealing with complex clients could generate this vicious circle of value depletion which in the long run may change the service nature from a partner-like relationship to a mere buyer–seller transaction.

The VDC is closely related to the concept of value co-destruction, which has been studied in the literature (Plé and Chumpitaz Cáceres 2010; Echeverri and Skålen 2011; Smith 2013), however the depletion circular process differs importantly in terms of the time frame. The VDC incorporates a *self-reinforcing circular process* due to the experiential phenomenon by which actors' dissatisfaction or negative experiences are repeated over time again and again. This repetitive phenomenon had not been observed by previous studies which look at value destruction as a series of one-off events, not a cumulative longer-term process.

Figure 7.1 depicts the process by which repeated low-value service experiences in the long term can generate detrimental consequences, not only for commercial companies like advertising agencies but for any provider organizations such as HEIs involved in unsuccessful service deliveries, and crucially also for the sector as a whole. Once this circle has commenced the service organization begins to be considered as simple providers by customers or users and it is difficult to recover their original status of being unique professional experts and partners in a mutually beneficial co-creation process. Customers often then begin to perceive the service to be low value, thus not co-creating value and contributing to the self-reinforcing circle of value depletion.

The VDC can be applied to the HE service ecosystem and in this context it operates as follows: we can identify examples of polluting practices in HE such as those related to student evaluation surveys or teachers' success rates which we have discussed earlier. Teachers will adapt their activities and behaviour to the demands of the system -institutional rules- by promoting positive feelings among their students

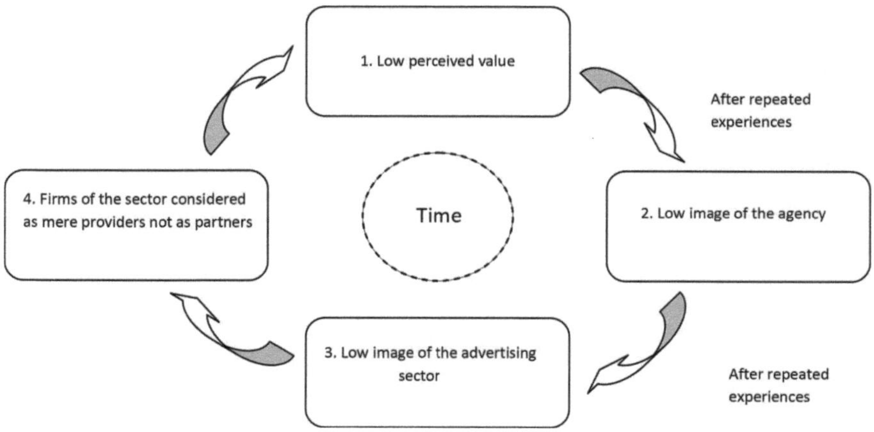

Fig. 7.1 The value depletion circle. *Source* Adapted from Díaz-Méndez and Saren (2019)

in order to get good evaluations from them. Satisfying students in the short term is relatively easy, but this alone, in the absence of other beneficial educational activities, will jeopardize the purpose of HE of training and instructing students to become good professionals and citizens in order to fulfill University's commitment to employers, society and humanity. Many teachers are constrained by the results of these surveys which can directly influence their professional reputation, university status and their salaries or other working conditions.

Over time this state of affairs will produce unprepared and frustrated graduates, and, on the other side, poorly respected, under-rewarded teachers and unsatisfied employers. Universities will be blamed by all sides and their role in society will start to be undervalued and even mocked with common remarks like "nowadays anyone can get a degree". This is not a trivial matter because the University was the pillar of societies' progress and civilization and its destruction could imply the de-evolution of society at a time when the distinction between fact and fiction, truth and lies, is disintegrating in many areas of public life. Who will we rely on to instruct our future professionals? This process which is taking place in many universities is depicted in Fig. 7.2.

In this chapter we argue that dramatic changes are ungently required in the institutional arrangements that result in the VDC scenario shown in Fig. 7.2 For example, student surveys can be undertaken but not to be used by university or State authorities as measure of teaching performance nor as a punishment/reward tool. They should only remain as one source of information for the teachers own pedagogic self-assessment, training and professional development. We advocate that a value co-creation approach should be adopted in order to save the University's essence and its critical role in society. By adopting this perspective, the educational process can become a collaborative process with joint efforts where both students and teachers

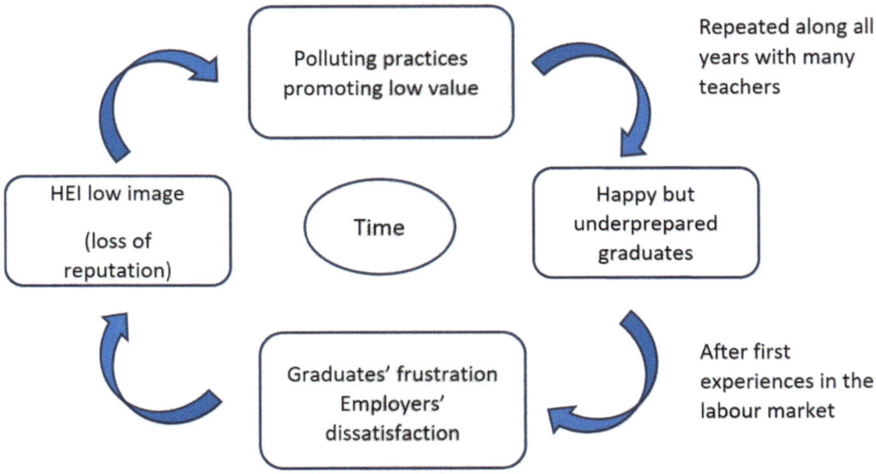

Fig. 7.2 VDC in HE. *Source* Own elaboration

share responsibility for HE service educational outcomes, fostering a more engaged and effective learning environment. This approach will allow teachers and students to focus on the real objective of HE, leaving aside factors that hinder (pollute) academic work.

This VDC can be interpreted as an amplifying feedback loop of the results of bad practices on value. In this sense, the SDL service ecosystems framework moves away from a linear, sequential flow of creation or destruction of value (Vargo et al. 2020) and points towards a more complex examination as explained here.

Value Co-Creation Approach to Counteract HE Service Ecosystem Pollution

In a strategic orientation that enables HEIs to face the crossroads of applying marketing strategies in a complex service ecosystem, researchers and academics are adopting a more reflective stance on market approach limitations and their impact on academia to avoid undermining their social value. In this regard, several developments in the literature on services emerged, such as Service Science (Maglio and Spohrer 2008), Many-to Many Marketing (Gummesson 2008, 2014), the Viable Systems Approach (Barile et al. 2016), Service Logic in Marketing (Grönroos 2006, 2011) and Service-Dominant Logic (SDL) (Vargo and Lusch 2004, 2008, 2017). These service developments provide an active student-centered approach and understanding of their role in academic activities (Díaz-Méndez et al. 2017) as co-creators of the value they can obtain from the HE service.

This chapter is framed on the SDL, service ecosystem perspective, which considers education is not a mere transaction between provider and customer, but an active collaboration in which the student plays a central role (Díaz-Méndez and Gummesson 2012; Gu and Wang 2022). This approach recognizes that the HE service ecosystem is made up of multiple actors, always including the student and their necessary interaction for the joint achievement of value during the educational service. SDL considers involving the student as a key factor in the achievement of academic objectives (Vargo et al. 2015). The idea that students are not mere recipients of knowledge, but actors who construct their knowledge, represents a significant change in the way marketing experts traditionally conceive interaction during educational service (Bovill 2020; Thomas and Ambrosini 2021).

Polluting practices, for the most part, originate from considering the student as a customer from the classical perspective of pioneering marketing. In this context, it is crucial to identify the differences between what is expected from the student's co-creation behavior from what is expected from the customer. Although both share some characteristics, the objectives they seek as value are different (Hemsley-Brown y Oplatka 2006). This distinction helps to prevent education from being reduced to a simple commercial transaction, focused solely on short-term satisfaction and immediate expectations, which undermines its intrinsic social value.

Students participation in value generation during their HE training and professional service involves students engaging in specialized, high-cognitive effort activities and endeavors that enrich their education (Dollinger et al. 2018). To better understand the role of the student as a co-creator of value in academic activities, the Student Value Cocreation Behavior (SVCB) framework can be considered (Zamora-Ramos et al. 2023). In this approach, students assume a central role as co-creators of their own educational experience, contributing to their individual needs, expectations and skills. By integrating their operational and operant resources in the classroom, motivation, participation and satisfaction can be increased. This co-creation approach does not neglect the pursuit of student satisfaction but recognizes that it must be managed differently.

The Role of the Value Co-Creator Student

Marketing management is shifting towards a service-based exchange logic that emphasizes the importance of understanding the context in which value is created (Permatasari et al. 2021). This approach challenges the traditional producer-customer relationship and advocates for more customer participation to generate value jointly. The HEIs that adopt co-creation do not claim to know the ideal needs of their students but instead prioritize maximizing student participation and interaction in teaching–learning processes to achieve more comprehensive outcome (Bovill 2020). This approach differs from HEIs that solely rely on their teachers to create value for students (Dollinger et al. 2018).

According to Vargo et al. (2008), marketing strategies that aim to create service-based exchanges rely on two interconnected elements: (a) the service itself, which involves utilizing the resources to benefit another party involved in the exchange, and (b) the value that is always co-created because of this exchange. In this sense, the service ecosystem perspective is the most appropriate way to understand this relationship, also proposing a recent concept of HE sector value, considering its unique characteristics.

Also, the value is subjective and can vary from student to student and from situation to situation (Akaka et al. 2013). This is relevant in the context of HE, as even if teachers use the same methods and tools, the outcome of student learning may differ (Díaz-Méndez and Gummesson 2012; Doyle et al. 2021). Additionally, based on the principles of SDL, value is phenomenological which implies that the student is the one who places value on their learning and the development of their skills for life and work. In this way, it is understood that when the students evaluate the quality of education based on the performance of their teachers and not on their own participation, the responsibility of the main actor is omitted.

Another characteristic of value is that it is multidimensional and created through coordination between actors and institutional arrangements. This means that the social context and culture in each university community influence the concept of value and how it is created. As a result, the value of HE is subjective and changes over time and in different environments (Ranjan and Read 2016). These social particularities and cultural changes are beyond the control of teachers. Although they can implement teaching strategies that help to integrate students and generate an optimal learning environment, they cannot control elements that are external and specific to the context of students.

The value co-creation approach in HE focuses on developing key skills in college graduates, such as critical thinking, effective reading and writing, complex problem solving and understanding diverse contexts (Soini et al. 2019). These cognitive and behavioral activities make up the SVCB, which includes co-design, co-production, and co-responsibility activities. These activities require a high level of quality communication and participation (Bovill 2020). Table 7.1 presents the characteristics that define a student as a co-creator in this ecosystem.

In this sense, the students are co-designers, co-producers and co-responsible in the HE service ecosystem, which is a partner in the process of learning design. To provide unique and interesting learning experiences that match the needs of each student, the involvement of the HE student is essential. Students may take control of their learning and contribute to their development and growth by working collaboratively with teachers and classmates during the service design process (Dollinger et al. 2018; Xu et al. 2018).

As collaborators in the educational process, students share the responsibility for directing that process and its results. These results may generate satisfaction or dissatisfaction, and in both cases, the student must assume his or her contribution. The idea of students as co-responsible in HE assumes a dynamic and participative

Table 7.1 Student as a co-creator of value

Co-designer	Co-producer	Co-responsible
Feedback Provide valuable comments to improve the learning experiences	*Specialized participation* Active integration in discussions and academic activities	*Service orientation* Understanding education is a dynamic interaction and adopting a service-oriented attitude
Collaboration Help find innovative solutions	*Self-taught* Self-directed learning and resource management	*Personal commitment* Assume the importance of their academic training and act accordingly
Reflection Analyze the growth and suggest improvements	*Experiential knowledge* By sharing knowledge and experiences	*Autonomy* To make decisions independently
Academic mature Make informed decisions about their education	*Practice-based learning* Embracing challenges and learning from failures	*Reflection* Continuous learning open-mindedness and adaptability

Source Own elaboration based on Dollinger et al. (2018)

function from the standpoint of SDL, where students are seen as co-creators and co-producers of value in the educational environment rather than as passive receivers of knowledge (Bovill 2020; Calma and Dickson-Deane 2020).

Students are active actors with distinctive knowledge, skills, and experiences that can be shared and used to benefit themselves and others through the lens of the SDL. This view encourages students to take charge of their education, actively seek opportunities for personal and professional development, and bring their knowledge and insights to the classroom to make it more interesting (Bovill 2020; Dollinger et al. 2018).

Theoretical Contributions and Limitations

This paper makes two principal original theoretical contributions to our understanding of the HE service ecosystem; i) introducing the concept of the value depletion circle (VDC) and applying it to the HE ecosystem and ii) problematization of current HE performance evaluation methods.

The theoretical concept and empirical phenomenon of the value depletion circle (VDC) is introduced as a central element in our analysis of the higher education service ecosystem. The VDC provides a causal explanation for the existence of several negative effects which we and other academics have identified as 'polluting' the HE ecosystem. We contend that our VDC analysis approach offers both a conceptual and managerial pathway to restoring balance within the educational ecosystem through the co-creation of value among all stakeholders, especially students, and to clearly define their roles as co-designers, co-producers, and co-responsible actors.

The core theoretical and practical problem inherent in the evaluation of HE quality that is highlighted and critically analysed in this paper is that the concept of value per se is difficult to define in the context of market exchange and therefore a *fortiore* the specification and measurement of value in education is an even greater challenge. HE institutions and regulatory bodies often adopt commercial evaluation practices appropriate for business stakeholders simply assuming that these also represent valid, comparable metrics of performance and quality of HE institutions. Our analysis and discussion of HE identified the key differentiating characteristics which affect the nature, creation and distribution of both value and quality compared to most private sector services. These features are i) the co-creation of value, ii) multiple inputs to quality that are not easily disaggregated, iii) many different types of beneficiaries and iv) supplier-specification of service quality which users find difficult to judge. We show how these attributes necessitate the careful design and tailoring of quality evaluation measures and methodologies that are appropriate for HE institutions.

There are two main limitations of this study. First, the transfer of the VDC to the HE context from the circumstances of our original identification and development concept in a commercial private sector market in a B2B context, in a study of the relationships between advertising agencies and their clients (Díaz-Méndez and Saren 2019), may require further refinement and elaboration than we have undertaken in this paper. Further empirical and developmental work is required to refine the application the VDC as an explanatory framework analytical tool to other public and service ecosystem phenomenological beyond that of HE.

Secondly, collectively the authors have extensive first-hand experience of international business school teaching and pedagogy, including evaluation methods in HEIs throughout the world. Nevertheless, the inevitable inescapable conscious and subliminal influences of their geographical and national backgrounds and experiences should be considered. Further similar VDC-based analytical studies in other countries, cultures and continents by researchers with local experience would be likely to add to our understanding of the nature of the problems and potential solutions more widely for HE teaching and evaluation methods that have been highlighted in this paper.

Conclusions

The introduction of beavers in Tierra del Fuego, Argentina, in the hope of replicating the benefits observed in Canada, resulted in an ecological disaster. This highlights the "Ecosystem Sensitivity" to show how one vital element of an ecosystem can destabilize another, and the importance of understanding the specific dynamics of each environment before introducing new species. Similarly, in the HE ecosystem, the adoption of quasi-mercantile practices, such as the evaluation of educational quality through student satisfaction surveys or the consideration of passing rates as an indicator of teachers' quality is generating adverse effects. These include grade inflation and the prioritization of short-term satisfaction over learning and skill development. This has led to a loss of credibility of HE among employers and society, threatening its fundamental role in human development as we have explained through the concept of the Value Depletion Circle which depicts the risk that polluting factors may have on the HE ecosystem. We have proposed the SDL and Service Ecosystem approach as a management approach to "rescue" this service ecosystem by offering an alternative way to restoring balance within the educational ecosystem through the co-creation of value among all stakeholders, especially students, and to clearly define their role as actors (co-designers, co-producers, and co-responsible). This approach seeks to counteract the negative effects of marketization by promoting integral development over immediate satisfaction. In this way, HEIs can re-assert themselves as key coordinators, facilitating the efficient interaction of operating and operational resources through institutional arrangements adapted to the dynamics and complexity of the ecosystem.

By viewing students not merely as passive consumers but as active co-creators of their educational experiences, HEIs can foster a more engaged and meaningful learning environment. This approach emphasizes collaboration, autonomy and continuous feedback, aligning educational outcomes more closely with student needs and societal demands. It encompasses not only immediate student satisfaction but also long-term outcomes such as career readiness and lifelong learning capabilities. Effective educational practices should account for this complexity, moving beyond simplistic metrics to assess the holistic development of students and the enduring impact of their educational experiences. Thus, the VDC and its negative impacts can be countered or blocked and thereby the role of university in society restored and preserved.

Addressing polluting practices in HE involves revisiting institutional arrangements and norms to ensure they support educational quality and student-centered outcomes rather than market-driven metrics alone. Abandoning exclusive reliance on student satisfaction surveys as indicators of educational quality is critical to restoring the integrity and effectiveness of the higher education system. Only then would be possible to ensure that HE truly fulfills its role as an enabler of human and social progress, preparing students not only for the present but for the challenges of the future with solid skills and a deep understanding of knowledge.

References

Akaka, M.A., Vargo, S.L., Lusch, R.F.: The complexity of context: a service ecosystems approach for international marketing. J. Int. Mark. **21**(4),1–20 (2013). https://doi.org/10.1509/jim.13.0032

Autio, E., Thomas, L.D.W.: Ecosystem value co-creation. Acad. Manag. Proc. **2018**(1), 15913 (2018). https://doi.org/10.5465/ambpp.2018.15913abstract

Baldini, A., Oltremari, J., Ramírez, M.: Impacto del castor (Castor canadensis, Rodentia) en bosques de lenga (Nothofagus pumilio) de Tierra del Fuego, Chile. Bosque (Valdivia) **29**(2), 162–169 (2008). https://doi.org/10.4067/S0717-92002008000200009

Banwait, K.S., Hancock, C.: Post browne review: a reflective analysis of marketisation dilemmas of senior managers in English universities. J. Mark. High. Educ. 1–18 (2021)

Barile, S., Lusch, R., Reynoso, J., Saviano, M., Spohrer, J.: Systems, networks, and ecosystems in service research. J. Serv. Manag. **27**(4), 652–674 (2016)

Berbegal-Mirabent, J., Ribeiro-Soriano, D.E.: Behind league tables and ranking systems. J. Serv. Theory Pract. **25**(3), 242–266 (2015). https://doi.org/10.1108/JSTP-04-2013-0059

Bovill, C.: Co-creation in learning and teaching: the case for a whole-class approach in higher education. High. Educ. **79**(6), 1023–1037 (2020). https://doi.org/10.1007/s10734-019-00453-w

Bovill, C., Cook-Sather, A., Felten, P.: Students as co-creators of teaching approaches, course design, and curricula: implications for academic developers. Int. J. Acad. Dev. **16**(2), 133–145 (2011). https://doi.org/10.1080/1360144X.2011.568690

Calma, A., Dickson-Deane, C.: The student as customer and quality in higher education. Int. J. Educ. Manag. **34**(8), 1221–1235 (2020). https://doi.org/10.1108/IJEM-03-2019-0093

Díaz-Méndez, M., Gummesson, E.: Value co-creation and university teaching quality: consequences for the European higher education area (EHEA). J. Serv. Manag. **23**(4), 571–592 (2012). https://doi.org/10.1108/09564231211260422

Díaz-Méndez, M., Saren, M.: Managing advertising agency client partnerships for value co-creation: characteristics, categories and challenges. Mark. Theory **19**(1), 9–26 (2019)

Díaz-Méndez, M., Paredes, M.R., Saren, M.: Improving society by improving education through service-dominant logic: reframing the role of students in higher education. Sustainability **11**, 5292 (2019). https://doi.org/10.3390/su11195292

Díaz-Méndez, M., Saren, M., Gummesson, E.: Considering pollution in the higher education (HE) service ecosystem: the role of students' evaluation surveys. TQM J **29**(6), 767–782 (2017). https://doi.org/10.1108/TQM-03-2017-0031

Díaz-Méndez, M., Saren, M., Gummesson, E.: Lights and shadows on student evaluation surveys: insights from service dominant logic. In: Gummesson, E., Díaz-Méndez, M., Saren, M. (eds.) Improving the Evaluation of Scholarly Work. Springer, Cham (2022). https://doi.org/10.1007/978-3-031-17662-3_6

Dollinger, M., Lodge, J., Coates, H.: Co-creation in higher education: towards a conceptual model. J. Mark. High. Educ. **28**(2), 210–231 (2018). https://doi.org/10.1080/08841241.2018.1466756

Doyle, E., Buckley, P., McCarthy, B.: The impact of content co-creation on academic achievement. Assess. Eval. High. Educ. **46**(3), 494–507 (2021). https://doi.org/10.1080/02602938.2020.1782832

Echeverri, P., Skålén, P.: Co-creation and co-destruction: a practice-theory based study of interactive value formation. Mark. Theory **11**(3), 351–373 (2011)

Elsharnouby, T.H.: Student co-creation behavior in higher education: the role of satisfaction with the university experience. J. Mark. High. Educ. **25**(2), 238–262 (2015). https://doi.org/10.1080/08841241.2015.1059919

Ferrary, M., Granovetter, M.: The role of venture capital firms in Silicon Valley's complex innovation network. Econ. Soc. **38**(2), 326–359 (2009). https://doi.org/10.1080/03085140902786827

Fleming, P.: The ghost university: academe from the ruins. Emancipations: J Crit. Soc. Anal. **1**(1), Article 4 (2021). https://scholarsjunction.msstate.edu/emancipations/vol1/iss1/4

Gao, W., Ding, X., Chen, R., Min, W.: An empirical study of the role of higher education in building a green economy. Sustainability **11**(23), 1–14 (2019). https://doi.org/10.3390/su11236823

Grönroos, C.: On defining marketing: finding a new roadmap for marketing. Mark. Theory **6**(4), 395–417 (2006). https://doi.org/10.1177/1470593106069930

Grönroos, C.: Value co-creation in service logic: a critical analysis. Mark. Theory **11**(3), 279–301 (2011). https://doi.org/10.1177/1470593111408177

Gu, W., Wang, J.: Research on index construction of sustainable entrepreneurship and its impact on economic growth. J. Bus. Res. **142**, 266–276 (2022). https://doi.org/10.1016/j.jbusres.2021.12.060

Gummesson, E.: Quality, service-dominant logic and many-to-many marketing. TQM J **20**(2), 143–153 (2008). https://doi.org/10.1108/17542730810857372

Gummesson, E.: Many-to-many marketing as grand theory: a Nordic school contribution. In: The Service-Dominant Logic of Marketing, pp. 357–371. Routledge (2014)

Hemsley-Brown, J., Oplatka, I.: Universities in a competitive global marketplace: a systematic review of the literature on higher education marketing. Int. J. Public Sect. Manag. **19**(4), 316–338 (2006). https://doi.org/10.1108/09513550610669176

Judson, K.M., Taylor, S.A.: Moving from marketization to marketing of higher education: the co-creation of value in higher education. High. Educ. Stud. **4**(1), 51–67 (2014). https://doi.org/10.5539/hes.v4n1p51

King, N., Bunce, L.: Academics' perceptions of students' motivation for learning and their own motivation for teaching in a marketized higher education context. Br. J. Educ. Psychol. **90**(3), 790–808 (2020)

Koskela-Huotari, K., Vargo, S.L.: Institutions as resource context. J. Serv. Theory Pract. **26**(2), 163–178 (2016). https://doi.org/10.1108/JSTP-09-2014-0190

Krstić, M., Filipe, J.A., Chavaglia, J.: Higher education as a determinant of the competitiveness and sustainable development of an economy. Sustainability **12**(16) (2020). https://doi.org/10.3390/su12166607

Maglio, P.P., Spohrer, J.: Fundamentals of service science. J. Acad. Mark. Sci. **36**, 18–20 (2008)

Marginson, S.: Putting 'public' back into the public university. Thesis Eleven **84**(1), 44–59 (2006). https://doi.org/10.1177/0725513606060519

Mason, C., Brown, R.: Entrepreneurial ecosystems and growth-oriented entrepreneurship. Final Report to OECD, Paris **30**(1), 77–102 (2014)

Nie, Z., Zurlo, F., Camussi, E., Annovazzi, C.: Service ecosystem design for improving the service sustainability: a case of career counselling services in the italian higher education institution. Sustainability **11**(5) (2019). https://doi.org/10.3390/su11051427

Permatasari, A., Dhewanto, W., Dellyana, D.: A proposed model of value co- creation through multi-stakeholder collaboration in domestic product development. Bus.: Theory Pract. **22**(2), 414–425 (2021). https://www.ceeol.com/search/article-detail?id=1007678

Plé, L., Chumpitaz Cáceres, R.: Not always co-creation: introducing interactional co-destruction of value in service-dominant logic. J. Serv. Mark. **24**(6), 430–437 (2010)

Polese, F., Sirianni, C.A., Grimaldi, M., Megaro, A.: (Co-)learning and (Co-) evaluation in scholarly ecosystem: challenges and opportunities in the Covid-19 era. In: Gummesson, E., Díaz-Méndez, M., Saren, M. (eds.) Improving the Evaluation of Scholarly Work. Springer, Cham (2022). https://doi.org/10.1007/978-3-031-17662-3_8

Quero, M.J., Ventura, R.: Balanced centricity in the higher education service ecosystem. In: Gummesson, E., Díaz-Méndez, M., Saren, M. (eds.) Improving the Evaluation of Scholarly Work. Springer, Cham (2022). https://doi.org/10.1007/978-3-031-17662-3_2

Ranjan, K.R., Read, S.: Value co-creation: concept and measurement. J. Acad. Mark. Sci. **44**(3), 290–315 (2016). https://doi.org/10.1007/s11747-014-0397-2

Readings, B.: The University in Ruins. Harvard University Press, Cambridge (1996)

Richter, C.H., Xu, J., Wilcox, B.A.: Opportunities and challenges of the ecosystem approach. Futures **67**, 40–51 (2015). https://doi.org/10.1016/j.futures.2014.12.002

Sánchez-Carrillo, J.C., Cadarso, M.A., Tobarra, M.A.: Embracing higher education leadership in sustainability: a systematic review. J. Clean. Prod. **298**, 126675 (2021). https://doi.org/10.1016/j.jclepro.2021.126675

Schelble, J.T.: The marketization of higher education in Hungary. Hung. Stud. **20**(2), 253–354 (2006). https://doi.org/10.1556/hstud.20.2006.2.6

Schiavini, A., Carranza, M.L., Deferrari, G., Escobar, J., Malmierca, L., Pietrek, A.G.: Erradicación de especies invasoras: ciencia, actitud y entendimiento. El castor en Tierra del Fuego. Mastozoología Neotrop. **23**(2), 279–288 (2016)

Seeman, E.D., O'Hara, M.: Customer relationship management in higher education: using information systems to improve the student-school relationship. Campus-Wide Inf. Syst. **23**(1), 24–34 (2006). https://doi.org/10.1108/10650740610639714

Sklyar, A., Kowalkowski, C., Tronvoll, B., Sörhammar, D.: Organizing for digital servitization: a service ecosystem perspective. J. Bus. Res. **104**, 450–460 (2019). https://doi.org/10.1016/j.jbusres.2019.02.012

Smith, A.M.: The value co-destruction process: a customer resource perspective. Eur. J. Mark. **47**(11/12), 1889–1909 (2013)

Soini, K., Korhonen-Kurki, K., Asikainen, H.: Transactional learning and sustainability co-creation in a university—business collaboration. Int. J. Sustain. High. Educ. **20**(6), 965–984 (2019). https://doi.org/10.1108/IJSHE-11-2018-0215

Thomas, L., Ambrosini, V.: The future role of the business school: a value cocreation perspective. Acad. Manag. Learn. Educ. **20**(2), 249–269 (2021). https://doi.org/10.5465/amle.2019.0239

Vargo, S.L., Lusch, R.F.: Evolving to a new dominant logic for marketing. J. Mark. **68**(1), 1–17 (2004). https://doi.org/10.1509/jmkg.68.1.1.24036

Vargo, S.L., Lusch, R.F.: Service-dominant logic: continuing the evolution. J. Acad. Mark. Sci. **36**, 1–10 (2008). https://doi.org/10.1007/s11747-007-0069-6

Vargo, S.L., Lusch, R.F.: Inversions of service-dominant logic. Mark. Theory **14**(3), 239–248 (2014). https://doi.org/10.1177/1470593114534339

Vargo, S.L., Lusch, R.F.: Service-dominant logic 2025. Int. J. Res. Mark. **34**(1), 46–67 (2017). https://doi.org/10.1016/j.ijresmar.2016.11.001

Vargo, S.L., Maglio, P.P., Akaka, M.A.: On value and value co-creation: a service systems and service logic perspective. Eur. Manag. J. **26**(3), 145–152 (2008). https://doi.org/10.1016/j.emj.2008.04.003

Vargo, S.L., Wieland, H., Akaka, M.A.: Innovation through institutionalization: a service ecosystems perspective. Ind. Mark. Manag. **44**(2013), 63–72 (2015). https://doi.org/10.1016/j.indmarman.2014.10.008

Vargo, S.L., Akaka, M.A., Wieland, H.: Rethinking the process of diffusion in innovation: a service-ecosystems and institutional perspective. J. Bus. Res. **116**, 526–534 (2020). https://doi.org/10.1016/j.jbusres.2020.01.038

Voropai, O., Pichyk, K., Chala, N.: Increasing competitiveness of higher education in Ukraine through value co-creation strategy (2019). https://doi.org/10.14254/2071-789X.2019/12-4/14

Xu, J. (Bill), Lo, A., Wu, J.: Are students customers? Tourism and hospitality students' evaluation of their higher education experience. J. Teach. Travel Tour. **18**(3), 236–258 (2018). https://doi.org/10.1080/15313220.2018.1463587

Zamora-Ramos, M.R., Chamorro-Mera, A., Díaz-Méndez, M.: Student value co-creation behaviour in the higher education service ecosystem: an empirical exploration. Market. Manag. Innovat. **14**(4), 140–150 (2023). https://doi.org/10.21272/mmi.2023.4-10

Montserrat Díaz-Méndez PhD is a senior lecturer in Marketing at the University of Extremadura, (Uex, Spain). She holds a PhD in Business Management from the University of Navarra. She is also the Director of the Business Management and Sociology Department at the UEx. She teaches and investigates on value co-creation, professional services and Higher Education ecosystems. She has published in numerous international journals.

Michael Saren PhD is a Professor of Marketing at Birmingham Business School, University of Birmingham. He is also an Emeritus Professor at Leicester University, where he was Head of the Marketing division 2004–2014. He was previously a Professor of Marketing at Strathclyde University and an Honorary Professor at St Andrews University. He has been a member of the academic Senate of the Chartered Institute of Marketing, the executive committee of EMAC, and was awarded an honorary fellowship and lifetime membership of the UK Academy of Marketing.

Moisés Rubén Zamora Ramos PhD is a Research Professor in the Marketing Department at the Autonomous University of the State of Hidalgo, with a PhD in Economics and Business and a Master′s in Marketing Management from the University of Extremadura, Spain. He also holds a Master′s in Management and Marketing from UPAEP, a Master′s in Education from UVEG, and a Bachelor′s in Foreign Trade from UAEH, along with specialized studies in Marketing Management at the University of Texas at Arlington, USA. His research focuses on Sustainable Marketing, SDL, and Value Co-creation.

Antonio Chamorro-Mera PhD is a Senior Lecturer at the University of Extremadura (Spain). He is a specialist in Strategic Marketing both in teaching and consulting. His research interests are focused on green and sustainable marketing, quality management and innovation. He has published in numerous international journals and currently he is the director of the University Institute for Research on Green Economy of the University of Extremadura.

Chapter 8
The Complexity of Interactions Between Automation and Human in Service Provisioning. The Role of Emotional Intelligence

Marisa Faggini, Marzia Del Prete, Bruna Bruno, and Anna Parziale

Abstract The role of EI in engaging customer during an automated service provisioning and in the subsequent value co-creation processes will be highlighted. In doing so, the contribution of AI in the recognition customers' emotions during automated service encounters will be analysed, focusing on the way by which firms should manage the increasing use of artificial intelligence (AI) in frontline interactions to sustain customer engagement, value co-creation if in customer service an AI technology detects customer emotions. In this context, the paper supports the logic of human–machine complement rather than substitution of human operators by AI (i.e. chatbots).

Keywords Value-co-creation · Emotional awareness · AI-technologies · Customer engagement

M. Faggini (✉) · B. Bruno
Associate Professor in Macroeconomics, Department of Economics and Statistics, University of Salerno, Salerno, Italy
e-mail: mfaggini@unisa.it

B. Bruno
e-mail: brbruna@unisa.it

A. Parziale
Fellow in Macroeconomics, Department of Economics and Statistics, University of Salerno, Salerno, Italy
e-mail: aparziale@unisa.it

M. Del Prete
AI & Data, Deloitte (Italy), Milano, Italy
e-mail: madelprete@deloitte.it

D. Sarno et al. (eds.), *Complexity and Emergence in Market Ecosystems*, New Economic Windows, https://doi.org/10.1007/978-3-031-81942-1_8

Introduction

A recent stream of research has highlighted the importance of customer engagement in service provisioning (Fonferek et al. 2019), even when service interactions are automated, digitally mediated and/or when they happen in a virtual services cape (Hollebeek et al. 2022; Torkzadeh et al. 2022). Some scholars pointed out the active role of advanced digital technologies (Zheng et al. 2019) in engaging customers (Simumba and Nchito 2018,) and in the co-creation on pleasant and unique experiences (Wirtz et al. 2018; Faggini et al. 2021). In this vein a relevant focus is addressed to analyse the influence (positive or not) that advanced and intelligent technologies can have on the whole customer journey intended as a multichannel/omnichannel and time-based process in which customers take part interacting with one or more services (Halvorsrud et al. 2016). This phenomenon has led companies to open towards an omnichannel (Bell et al. 2014; Verhoef et al., 2015) and automated approach to service, as basic conditions for offering a better emotional experience to customers (Nyrhinen et al. 2022).

Nevertheless, often technology-mediated interactions as well as the absence of the so-called "human touch" can negatively affect customer experience and her/his psychological disposition to be engaged into service provision (Nicolescu and Tudorache 2022). This, for example, happens when consumers interacting with AI application, such as a chatbot, feel discomfort (Luo et al., 2019), un-satisfaction and/or frustration due to misinterpretation of their requests (Forbes 2019), producing dis-engagement and emphasizing the inability of most chatbots to handle the complexities of human conversation (Rizomyliotis et al. 2022).

Provided that strong relationships, high quality communication and sustained dialogue are necessary to co-create value during a customer-chatbot interaction (Jaworsky and Kohli 2006, Prahalad and Ramaswamy 2004; Auh et al. 2007), how can virtual operators work to completely understand customers' emotions and engage them during their journey without losing the emotional connectedness? One possible answer, we try to explore through this paper, is to combine the skills of Emotional Intelligence (Goleman 1998) with those of current AI-powered chatbots.

In traditional service encounters, emotions can be tuned through emotion regulation by Front Line Employees to achieve desired outcomes (Puccinelli et al. 2010; Tamir 2016). Moreover, according to their own personality customers, attempting to resolve their dissatisfaction with services may hide negative expressions and present pleasant demeanours, thus engaging in acting strategies (Ligas and Coulter 2001). Alternatively, they might engage in deep or surface acting to modify felt emotions in response to the human operator's behaviour (Medler-Liraz and Yagil 2013). Such behaviours demonstrate that customers understand the emotions that are appropriate to show or hide in service encounters. Enhancing the understanding of the role of emotions and personality in consumer behaviour means that customer-chatbot interactions can be more effectively tailored to respond to customer needs and better matching products and services, reducing search costs (Matz et al. 2017).

In particular, the reciprocal nature of emotional behaviour in service encounters demonstrates the importance of understanding the concepts of emotional awareness (EA), as component of Emotional Intelligence (EI), for putting it at the service of artificial intelligence tools such as chatbots as well. It follows that a high-quality servicescape "can encourage cognitive and affective psychological states that beget customer engagement" (Del Prete 2021; Torkzadeh et al. 2022 p.10), acting on her/his emotions, satisfaction, and behavioural intentions related to provided product/services. This also represent one of the most recent drivers that firms can use for taking a step forward differentiation and to achieve sustainable competitive advantage (Sun et al. 2020).

However, few research examined how to design chatbots capable of understanding and addressing consumers 'emotional distress, particularly in situations involving product or service failures (e.g. Chong et al. 2021; Tsai et al. 2021). Even if more attention is called for improving the cognitive and behavioural dimensions of customer experience (Puntoni et al. 2021), much less it has been devoted to investigate the role of emotional intelligence in automated customer service, where the AI-powered chatbots is used as a complement or substitute for human frontline employees (Lajante et al. 2023; Ruan and Mezei 2022).

Drawing on these considerations, the role of EI in engaging customer during service provisioning and in the subsequent value co-creation processes will be highlighted. In doing so, the contribution of AI in the recognition customers' emotions during automated service encounters will be analysed, focusing on chatbots readiness or not in perceiving human emotions all-over the customer journey. In this context, the paper supports the logic of human–machine complement rather than substitution of human operators by chatbots (Huang and Rust 2018; Tschang and Almirall 2021).

Therefore, the chapter has been organized as follows. In paragraph 1 how EI can be used as strategy for customer engagement will be described. The second investigates how AI fits into customer service systems through its own conversational agents (chatbots), activating an inevitable change. The third describes the conceptual framework of emotional awareness in automated service encounters for chatbots and analyse the impact on value co-creation with respect to interactions with human operators. Discussion and implications of the work are highlighted in the final paragraph.

Emotional Intelligence as Customer Engagement Strategy in Service Provisioning

Customer engagement (CE), foundational element of value co-creation (Storbacka et al. 2016), is a complex and multifaceted concept, which the literature has differently approached. Generally, it has been considered as the degree to which customers are involved in creating and providing services (Dong and Sivakumar 2015; Zhao et al. 2018). Even though the literature offers several different definitions of CE, all of them tend to emphasize its behavioural and emotional focus (van Doorn et al. 2010; Hollebeek 2011; Vivek et al. 2012), involving the different actors/stakeholders (e.g., customer/client, employees, suppliers, partners, etc.). Emotions impact on their and they are crucial in interaction among them.

To be emotionally engaged with a firm means that firm aligns with personal motivations of customer enabling them to fulfil deep desires, some of which they are not even aware of (Scott et al. 2015). However, sometimes customers can refuse to be engaged in service provisioning for some different reasons, such as the loss of emotional connectedness and the emotional "extraneity" to a brand (Khuhro et al. 2017).

Emotional Intelligence is shown to play an important role for managing emotions during service encounters in many sectors (Danquad and Wireko 2014) to provide much better service and have more satisfied customers. Some scholars (Wolfe and Kim 2013) demonstrated that services providers with a higher level of EI tend to perform better and to receive less complaints thanks to their ability in managing emotions during the service encounter. Consequently, this implies a gain in customer's loyalty and advocacy and in firm performance (De Keyser et al. 2015; Zaid and Patwayati 2021).

Salovey and Mayer (1990) introduced EI describing it as "the subset of social intelligence that involves the ability to monitor one's own and other's feelings and emotions", as well as, "the ability to discriminate among them and to use this information to guide one's thinking and actions" (Hein 2007). They further described emotional intelligence as, "a set of skills hypothesized to contribute to the accurate appraisal and expression of emotion in oneself and in others, the effective regulation of emotion in self and others, and the use of feelings to motivate, plan, and achieve in one's life" (Hein 2007). In the 1995, Goleman arguing that emotional competences are not innate talents, but they can be learned (people can be trained to acquire emotional intelligence) developed a model of EI consisting of five main constructs depicted in Table 8.1.

Following this statement emotionally intelligent service providers may have cognition of:

- themselves and their emotions;
- other people and their feelings and what signal such feelings give off;
- the impact they have on others;

Table 8.1 Main constructs of EI by Goleman (1995)

Self-awareness	The ability to recognize your own feelings and how they affect others
Self-regulation	The ability to control or redirect emotional impulses, especially those that are negative
Motivation	A passion to work for reasons other than money
Empathy	Recognizing others' feelings and responding appropriately
Social skills	The ability to manage relationships and build social networks

- the impact other people have on them.

Therefore, with this knowledge, they will be able to manage difficult conversations with the customer (Goleman 2001). It follows that the implementation of emotional-based strategies can increase the percentage of customer engagement and reduce the percentage of customer attrition. Moreover, these strategies can positively affect the strength and the direction of CE throughout the customer journey because it is mainly influenced by the emotional connectedness at the core of a socio-economic interaction, which happens in service encounter. In fact, they can contribute to solve emergent problems and establish a better emotional connectedness with customers creating an intimate and empathic relationship (Pansari and Kumar 2017).

In this vein EI (Goleman 1998, 2001) offers an essential contribution to CE positively affecting the relationships (Lopes et al. 2004, 2005) between customers and firms' physical and/or digital touchpoints (Kenbach and Nicola 2005, Radha and Prasad 2013).

Moreover, this construct has been also considered as a potential strategy for retaining and involving customer in service provisioning (Wirtz et al. 2013) also through cognitive, emotional, and behavioural presence in brand interactions with an online community and/or a particular online shopping platform (Hollebeek 2011). This happens when these platforms offer a good and pleasant servicescape, which makes customer able to live positive experience and to participate in different stage of service provision. By involving the customers in service production, providers can capture customers' needs and maintain their competitiveness. Even if creating an engaging servicescape usually need for a certain initial investment, however it can potentially boost firms' performance in terms of sales (Neff 2007) and profits growth (Verhoef et al. 2010) as well as in terms of the long-lasting competitive advantage (Sedley 2008).

Artificial Intelligence for Omnichannel Services: The Role of Chatbots in Service Encounter

The ever-changing retailing practices and consumers' behaviour together with the global technological advancement has paved the way for disruptive transition (Makridakis 2017) towards an omnichannel service experience (Gerea et al. 2022). This evolution led to redesign also the customer journey, making it inherently customer-centric (Heinonen and Strandvik 2015). It follows that customers are currently able to seamlessly and interchangeably move among multiple channels to search for information about products and services, sharing opinions and experiences, and talking about and with a company (Piotrowicz and Cuthbertson 2014).

Differently from a multichannel approach where the channels are not necessarily integrated with each other, the omnichannel approach aims at creating a seamless experience across touchpoints (Fig. 8.1) in the customer lifecycle—websites, social

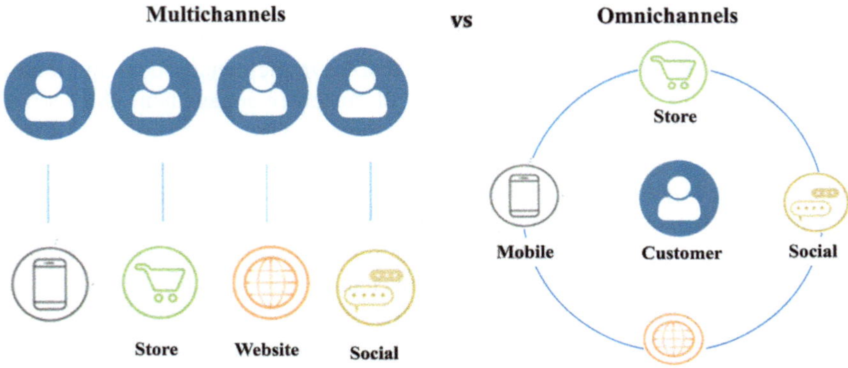

Fig. 8.1 Multichannel versus Omnichannel customer journey. *Source* authors' elaboration

media, phone calls, and store both online and offline, offering to consumers a consistent, uniform, and integrated service and experience regardless of the channel they use.

Almost all customers demand customer care serice available 24 h a day, seven days a week, as their wants and preferences change frequently.

In omnichannel customer journey, firms prioritize being online anytime and anywhere to keep in touch with customers and identify the easiest, need-relevant, and pleasurable channel for engaging customers and managing its automated customer journey (Grewal et al. 2017). Handling this by a human agent is quite unlikely.

Therefore, firms are leveraging new technologies from AI and one of the most recent and increasingly popular AI-based applications in customer service are chatbots.

Chatbot, or disembodied conversational agents (Bolton et al. 2018) frequently are used in current customer service systems (Kvale et al. 2020), providing a viable, automated solutions to customers more quickly than human agents (Nguyen et al. 2022). They are AI-powered algorithms and dialogical systems that simulates an interactive human conversation by using key pre-calculated user phrases and auditory or text-based signals, giving to consumers the feeling of chatting with a real person and allowing a real-time dialogue. The use of technology-mediated services allows customers to derive various benefits (Rzepka et al. 2021), like instant assistance anytime and anywhere, provision of links to obtain relevant information, resolution of problems with the aid of text or symbols (Li and Zhang 2023).

Already in the 2019, the CapGemini report (Taylor et al. 2019) showed the growing trend in chatbot use that today is steadying, arguing that in retail banking and insurance, 49% of the top 100 organizations employed chatbots while in consumer products and retail, the corresponding number was 23%. This growing chatbot implement was motivated by improved language processing capabilities due to advances in artificial intelligence and Natural Language Processing (NLP) technologies but also by customer service costs reduction and interactions increasing with customers more than 20% (Fig. 8.2), and improving customer experience their satisfaction level increases (Fig. 8.3).

Nevertheless, it is worth noting that a customer's preference for interaction with chatbots vs human employees depends on the following characteristics, amongst others (Dabholkar 1996, Simon and Usunier 2007):

- Rational (economic efficiency), optimistic, innovative, and technologically ready customers prefer a chatbot interaction.

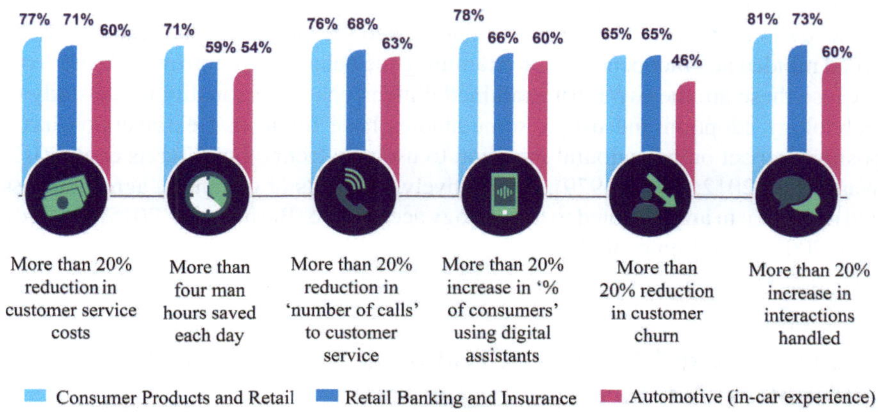

Fig. 8.2 Benefits from using chat assistants. *Source* Capgemini Research Institute Report 2019

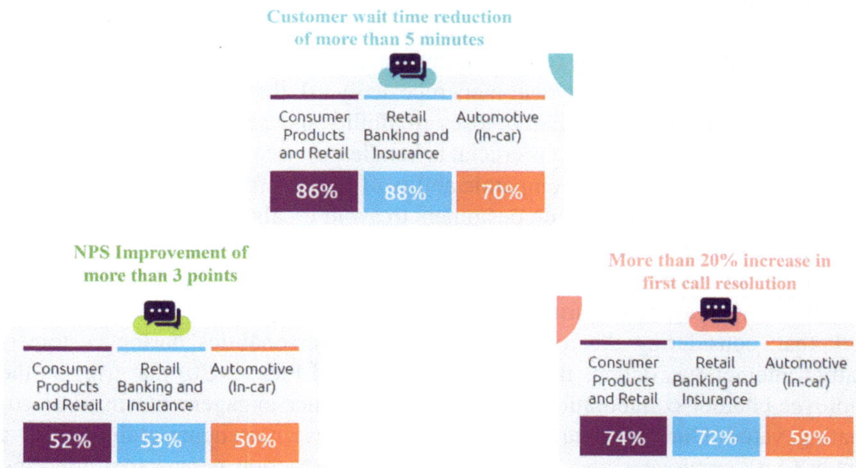

Fig. 8.3 Customer satisfaction. Authors' elaboration from Capgemini Research Institute Report 2019. NPS: Net Promoter Score® (NPS)

- Emotional, dependent on personal recommendation, and technologically anxious customers prefer an interaction with human staff.

Rational vs emotional customer has different approaches during an interaction with a chatbot and being customer-chatbot interaction a form of communication it may be enhanced based on personality attributes (Moon 2002). The five-factor model (FFM) or Big Five is the most comprehensive taxonomy used in research for personality studies (McCrae and Costa 1987). According to John (1990), FFM traits of personality are: (1) Agreeableness: the tendency to be compassionate, tolerant and cooperative toward others; (2) Conscientiousness: the tendency to act in an organized manner; (3) Extraversion that is the tendency to socialization; (4) Neuroticism: the tendency to be nervous, insecure or anxious; and (5) Openness: the tendency to be broad minded and independent. On examining the behaviour of introverts-extraverts, because these are the two most examined dimensions of personality in the study of technology adoption and usage, some studies have found that extraversion has a positive impact on behavioural intention to use new technology (Terzis et al. 2012; Wang et al. 2012; Zmud 1979). Alternatively, the traits of openness, agreeableness and neuroticism are unrelated to technology acceptance (Barnett et al. 2015; McElroy et al. 2007; Svendsen et al. 2013).

Engagement and Value Co-Creation in Automated Service Encounters: A Conceptual Framework

In automated service encounters a lack of empathy or an inability to connect emotionally with customers due to emotional limitations of AI-powered chatbots might dramatically damage customer engagement over time (Agnihotri and Krush 2015).

When a chatbot completely replaces human staff, it could be possible to co-destroy the value (Čaić et al. 2018). If customers repeatedly ask the same questions without getting relevant answers, the chatbot loses credibility and customers will soon break off the communication. When it is crucial to handle a service disruption or respond to customer service requests, the empathic ability of FLEs is functional in establishing an emotional connectedness with customers to avoid escalation toward loss of trust in companies (Wieseke et al. 2012).

With the aim to provide comprehensive and practical guidance for practitioners as well as encouraging empirical future research, we propose a conceptual framework that, considering the role of emotions and personalities during customer-chatbot interactions, defines the SP (Lajante and Del Prete 2020) that triggers the employee-chatbot collaboration for a perfect customer engagement aimed at co-creating value. Our conceptual framework identifies antecedents mainly from a human (e.g., employees and customers)-chatbot interaction perspective, ignoring the external and environmental factors that possibly affect the emotion process.

The proposed conceptual framework, as shown in Fig. 8.4, is organizing in the following manner. First, we discuss the concept of the SP in chatbot-customer interaction. Then, we identify a critical antecedent of the SP, namely emotional awareness (EA), the first component of EI. Since service is a co-creation process with both a service provider (firm) and service recipient (customer), we discuss the antecedent EA from the perspective of both firms and customer. Finally, we discuss the consequences of the SP that affects firm performance through an increase in the perceived quality of service and service experience. In our analysis CE is represented by the links between satisfaction and direct customer engagement and between emotion and indirect customer engagement as well as addressed in the literature (Pansari and Kumar 2017).

The conceptual framework starts from a customer interaction with a chatbot (antecedents). The chatbot through an EA plug-in can recognize whether the personality and the (positive/negative) valence of the emotion of the customer requires the presence of a FLE or not. Indeed, a chatbot equipped with an EA algorithm could detect the customer's personality and emotion from the text at the beginning of the interaction and adjust the transitions between emotions during the conversation thanks to the SP. The SP towards a FLE occurs when the chatbot with EA recognizes a negative emotion capable of interrupting the conversation and therefore associates it with a complex task to be diverted to FLE. LEs will then be able to use empathy

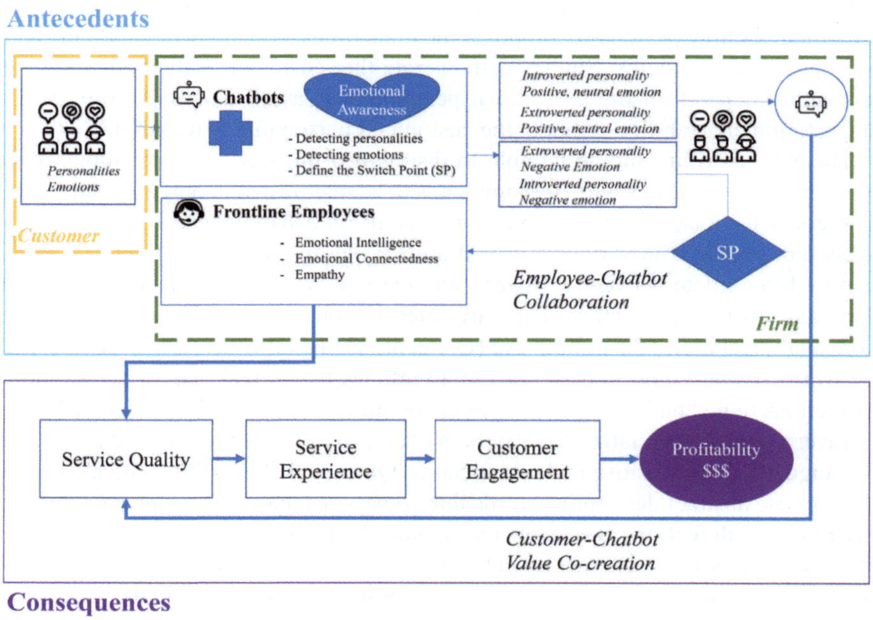

Fig. 8.4 Conceptual framework. *Source* authors' elaboration

and their emotional skills to positively resolve the customer's request and sustain CE during the conversation.

When the chatbot recognizes a positive or neutral emotion could manage the task autonomously and engage the customer in a pleasant conversation by assuming personalities and moods like those of the customer through his own capacity for emotion regulation (Lee et al. 2014, Lee et al. 2020b).

The emotion identification process of EA starts with the two stages of *detection* and *regulation* and is a prerequisite for a chatbot to interact with customers. Although emotion identification is automatic for humans, it is still a challenge for chatbots. This process needs different elements (hardware and software) to detect personalities and emotions for understanding human emotional behaviour. For instance, sensors' data might feed the chatbot database with emotion-related physiological information. Afterward, applying an EA algorithm in interspecific service encounters (Robinson et al. 2020) would help to analyse and use the emotional content to connect with customers and recognize their emotions at the start of the interaction.

Therefore, from our point of view the EA algorithm should have three components:

1 personalities detection;
2 emotions detection;
3 emotion regulation through the definition of the SP.

Personalities detection. It is well known that utterances convey a lot of information about a speaker in addition to their semantic content of the text. Computational work on modelling personality proposes methods for expressing personality in chatbots and tutorial systems. Studies demonstrate that customer evaluations of agent personality depend on the user's own personality (Cassell and Bickmore 2003), suggesting that an ability to model the customer's personality is required.

The EA algorithm should be able to distinguish introverted personalities from extroverted personalities. Extroversion tends to be manifested in outgoing, talkative, energetic behaviour, whereas introversion is manifested in more reserved and solitary behaviour. People with various personal traits write differently and linguistic signs help to differentiate between introverts and extroverts. Introverts spend more time alone and avoid socialization. They are often in solitary activities and tend to be more analytical rather than impulsive (Lee et al. 2014). Introverts are more likely to use words that are more precise, use more articles (e.g., "the", "a") and use more distinctions (e.g., "but", "however") (Beukeboom et al. 2013; Schwartz et al. 2013). Extraverts prefer to socialize (Beukeboom et al. 2013). They are more talkative and use language that is positive and self-focused (Qiuet et al. 2012). They are also more likely to use abstract descriptive words that sometimes may require interpretation.

Emotions detection. Based on the appraisal theories (Ellsworth and Scherer 2003), emotions are states that reflect evaluative judgments (appraisal) of the environment, the self and other social entities, considering the individuals' goals and beliefs, which motivate and coordinate adaptive behaviour (Hudlicka 2016). One of the most used sources for emotionally labelled text is distributed by the Swiss Centre for Affective Science. It consists of responses from about three thousand people worldwide who were asked to report situations in which they experience each of

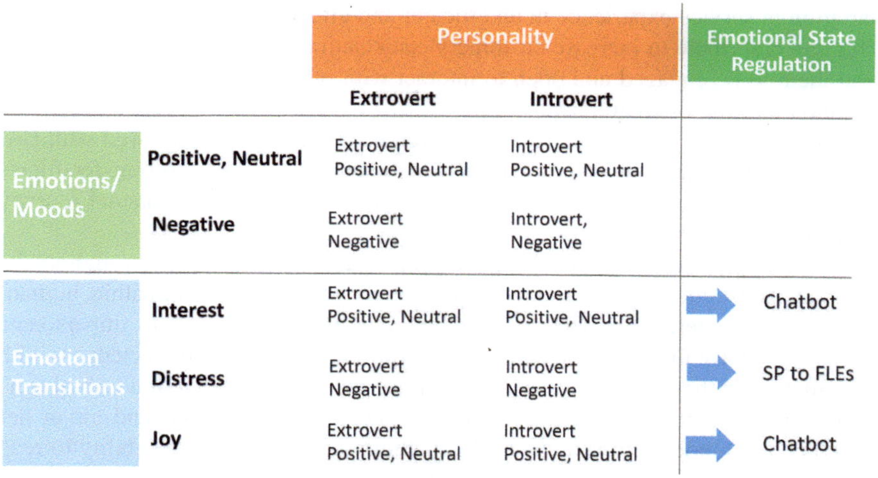

Fig. 8.5 The EA matrix. *Source* authors' elaboration

these seven emotions: joy, fear, sadness, disgust, shame and guilt and how they react to them. This dataset consists of about 7600 records of emotion-provoking text.

The topic of the transitions between emotions also is fundamental to detect present and future emotions. Picard (1997) uses the state transition diagram to simulate transitions between emotions. Interest (I), distress (D) or joy (J) characterise the three "hidden" states of a person. For example, if a customer is interested in having information about a product but he is unable to obtain it, he transitions into distress. Once information is obtained, he can further transition to joy. Successful social interaction depends on people's ability to predict others' future actions and emotions. For a chatbot is important to detect both emotions and emotion transitions. The matrix in Fig. 8.5 shows how the EA algorithm works.

Emotion regulation. It represents the process people use to monitor, evaluate and modify emotional responses, including emotional experiences and expression (Gross et al. 2006).

In our conceptual framework emotion regulation is the capability of the chatbot with an EA plug-in to detect the SP in case of negative emotion and customer distress or to satisfy the customer's request in case of positive emotion, interest or joy.

Recognising customers managed emotional displays, thanks to effective chatbot-FLEs integration is important for providers, as they try to provide the emotional feedback customers require. An EA-powered chatbot represents a good example of effective integration of resources in managing service interaction with customers in an overall vision of co-creation produced by both chatbot-FLE collaboration and productive chatbot-customer interaction (Fig. 8.4). Ensuring consistency in service offering is very challenging because service performance of FLEs may vary at different point in time and customers prefer a consistent service experience (Parasuraman et al. 1991; Wang et al. 2018). Chatbots with EA could help firm reduce the variation in service experience in an efficient and effective way, increasing the CE. This can contribute to increase customers' satisfaction with the service, which can lead them to be engaged and open to interact with other actors/stakeholder for the common purpose of creating value.

However, value co-destruction can be avoided when chatbot, endowed with EA, identifying the related switch point (SP) to a human touchpoint. This transfer allows for preventing a loss of emotional connectedness along the omnichannel journey (Lajante and Del Prete 2020).

The chatbots trained by EA monitor customers' emotional episodes throughout the automated customer journey and redirect them, if necessary, to an empathic, human-based service interaction. This implies that chatbots do not need to impersonate humans and have the empathic ability to establish an emotional connectedness with customers. Chatbots should be able to evaluate the service interaction in real time to recognize when there is a variation in the customer's emotions and his or her issue becomes more complex. As such, chatbots should develop the ability to read emotions and personalities to identify the exact point at which the conversation must be managed by a human agent, will preventing a loss of emotional connectedness and value co-destruction in the omnichannel customer journey. Examining the emotional content of the customer's experience can enhance the understanding of customer satisfaction and contribute to value co-creation.

Conclusions

With the spread of AI and the use of conversational robots as chatbots, firms can lose their empathic capacity (Lajante 2019) with consequent co-destruction of value. This problem could be avoided if the chatbot is used for simple questions to resolve immediately. For much more complex problems, often associated with negative emotions, it should be opportune to use FLEs.

Even if the chatbot adoption, and in generally of AI technologies, is justified by of cost- and time-saving opportunities, service availability and accessibility, however a chatbot is not an emotional bot that could manage difficult conversations as consequences of customers' negative emotions. Chatbots today do not have an advanced emotional reading (such as EA) of customer feelings, nor an empathic ability to motivate and engage him. Thus, this technology sometimes cannot really service consumers without human intervention. Customers expect that human operators will continue to play an essential role in defining frontline service encounters in a variety of service settings (De Keyser et al. 2019).

For this reason, we argued that there should be no substitution between AI operator and FLEs but rather collaboration aimed at value co-creation. The co-created value between FLEs and chatbots endowed with EA offer valuable digital opportunities for servitization to create and capture new value. Our conceptual framework is relevant in understanding the democratization of AI capabilities through the collaborative logics AI-FLEs.

A service encounter is typically interactive and reciprocal between customers and providers. The reciprocal nature of emotional behaviour in service encounters demonstrates the importance of understanding the concepts of EA and emotion regulation from the customer viewpoint. Our conceptual framework highlights how a customer-centric co-creation approach could be realized to deliver proof of value for chatbots solutions. The goal is on supporting CE with an initial clear value proposition that is subsequently refined and implemented in collaboration with the customer and FLEs.

Therefore, knowing when a human operator is required to manage a difficult interaction appears to be fundamental. The service providers must identify customers' negative emotional episodes and need for assistance to switch from AI-powered chatbots to human customer service at the right time. Finding this switch point, in the case of a difficult emotional conversation, means having more symbiotic interactions, engaging customers in a collaborative way, and avoiding the loss of the firm's empathic capacity (Lajante 2019). Therefore, firms should not shift towards customer automatic channels completely, especially not at the beginning of a relationship with a customer.

Deploying AI technologies endowed with EI strategically at right time may therefore bring significant benefits to firms and to increase CE. AI technologies adoption achieving significant productivity improvements allows to firm to be competitive on the market but a totally human substitution could set up a reverse mechanism: loss of that competitive advantage.

The proposed conceptual framework includes a new construct: the need for interaction in automated service encounter with a service employee to solve the lack of empathy of chatbot and to ensure customer engagement and satisfaction. A fruitful human–chatbot interaction becomes crucial to provide a better automated customer experience, to enhance customer satisfaction, and to increase profits. Therefore, automated customer journey, with included EI becomes an increasingly important part of a business strategy allowing for a better balance between growth and personalization.

More research in this area is needed because today chatbots are a frequent AI-tool in customer service and customers' emotional distress, need for assistance, and critical/sensitive periods (e.g., anxiety, frustration, lack of trust) are serious issues for firms who want to succeed in connecting emotionally and empathize with customers to enhance the customer experience and co-create value.

Knowing the customer's feelings and emotions at every touchpoint can make a real difference because as stated customer emotional connection established with the firm appears fundamental to create loyalty and advocacy, and then profits. Therefore, firms must make considerable investments in the development of machine interfaces without forgetting that among the distinctive features of human race there are the ability to feel emotions and to be empathetic with others.

The goal of incorporating EI dimensions in AI-powered chatbots is built an inter-human service encounter that contribute to value co-creation. It is worth noting which emphasis on this point where personalized consumer experience, also in AI era, takes central role as opposed to a product and firm-centric view, reshape the way firms are created, survive and grow.

References

Agnihotri, R., Krush, M.T.: Salesperson empathy, ethical behaviors, and sales performance: the moderating role of trust in one's manager. J. Pers. Sell. & Sales Manag. **35**(2), 164–174 (2015)

Auh, S., Bell, S.J., McLeod, C.S., Shih, E.: Co-production and customer loyalty in financial services. J. Retail. **83**(3), 359–370 (2007)

Bell, D.R., Gallino, S., Moreno, A.: How to win in an Omnichannel world. MIT Sloan Manag. Rev. **56**(1), 14 (2014)

Bolton C., Machová, V., Mišanková, M., Valášková, K.: The power of Human–Machine collaboration: artificial intelligence, business automation, and the smart economy. Econ. Manag. Financ. Mark. **13**(4), 51 (2018)

Čaić, M., Odekerken-Schröder, G., Mahr, D.: Service robots: value co-creation and co-destruction in elderly care networks. J. Serv. Manag. **29**(2), 178–205 (2018)

Cassell, J., Bickmore, T.: Negotiated collusion: modeling social language and its relationship effects in intelligent agents. User Model. User-Adap. Inter. **13**, 89–132 (2003)

Danquah, E., Wireko, T.B.: The impact of each element of emotional intelligence on customer service delivery: a customer satisfaction perspective. Int. J. Sales Market. **4**(2), 9–20 (2014)

De Keyser, A., Köcher, S., Alkire, L., Verbeeck, C., Kandampully, J.: Frontline service technology infusion: conceptual archetypes and future research directions. J. Serv. Manag. **30**(1), 156–183 (2019)

Del Prete, M.: Emotional artificial intelligence: detecting and managing customer emotions in automated customer service, Ph. D. Thesis, (2021). https://doi.org/10.14273/unisa-4563

Dong, B., Sivakumar, K.: A process-output classification for customer participation in services. J. Serv. Manag. **26**(5), 726–750 (2015)

Ellsworth, P.C., Scherer, K.R.: Appraisal processes in emotion. In: Davidson, R.J., Scherer, K.R., Goldsmith, H.H. (eds.) Handbook of Affective Sciences, pp. 572–595. Oxford University Press (2003)

Faggini, M., Bruno, B., Parziale, A.: Creating value for a sustainable healthcare: the role of digital platforms. J. Creat. Value **7**(2), 170–182 (2021)

Fonferek, K.-M., Kleinaltenkamp, M., Möller, J.: Customer engagement in the process of service provision. J. Serv. Manag. Res. **3**(1), 12–23 (2019)

Forbes (2019). Chatbots a powerful weapon in the business. https://www.forbes.com/sites/forbestechcouncil/2018/08/29/chatbots-a-powerful-weapon-in-the-business-arsenal/#41580cf64960

Goleman, D.: Emotional Intelligence. Bantam Books, New York (1995)

Goleman, D.: Working with Emotional Intelligence. Bantam Books, New York (1998)

Goleman, D.: An EI-based theory of performance. In: Cherniss, C., Goleman, D. (eds.) The Emotionally Intelligent Workplace, pp. 27–44. Jossey-Bass, San Francisco, CA (2001)

Grewal, D., Roggeveen, A.L., Sisodia, R., Nordfält, J.: Enhancing customer engagement through consciousness. J. Retail. **93**(1), 55–64 (2017)

Gross, J.J., Richards, J.M., John, O.P.: Emotion regulation in everyday life. In: Snyder, D.K., Simpson, J.A., Hughes, J.N. (eds.) Emotion Regulation in Couples and Families: Pathways to Dysfunction and Health, pp. 13–35. American Psychological Association, Washington DC (2006)

Halvorsrud, R., Kvale, K., Følstad, A.: Improving service quality through customer journey analysis. J. Serv. Theory Pract. **26**, 840–867 (2016)

Heinonen, K., Strandvik, T.: Customer-dominant logic: foundations and implications. J. Serv. Mark. **29**(6/7), 472–484 (2015)

Huang, M., Rust, R.T.: Artificial intelligence in service. J. Serv. Res. **21**(2), 155–172 (2018)

John, O.P.: The "Big Five" factor taxonomy: dimensions of personality in the natural language and in questionnaires. In: Pervin, L.A. (ed.) Handbook of personality: theory and research, pp. 66–100. The Guilford Press (1990)

Lajante, M., Del Prete, M., Sasseville, B., Rouleau, G., Gagnon, M., Pelletier, N.: Empathy training for service employees: a mixed-methods systematic review. PLoS One **18**(8), e0289793 (2023)

Lee, E., Ahn, J., Kim, Y.J.: Personality traits and self-presentation at Facebook. Pers. Individ. Differ. **69**, 162–167 (2014)
Li, C., Zhang, J.: Chatbots or me? Consumers' switching between human agents and conversational agents. J. Retail. Consum. Serv. (2023)
Ligas, M., Coulter, R.A.: Changing faces in services relationships: customers' roles during dissatisfactory service encounters. Adv. Consum. Res. **28**, 329–334 (2001)
Lopes, P.N., Brackett, M.A., Nezlek, J., Schütz, A., Sellin, I., Salovey, P.: Emotional intelligence and social interaction. Pers. Soc. Psychol. Bull. **30**, 1018–1034 (2004)
Lopes, P.N., Salovey, P., Côté, S., Beers, M.: Emotion regulation ability and the quality of social interaction. Emotion **5**, 113–118 (2005)
Luo, X., Siliang, T., Zheng, F., Qu, Z.: Frontiers: machines versus humans: the impact of artificial intelligence chatbot disclosure on customer purchases. Market. Sci. **38**(6), 913–1084 (2019)
Makridakis, S.: The forthcoming artificial intelligence (AI) revolution: its impact on society and firms. Futures **90**, 46–60 (2017)
Matz, S., Kosinski, M., Nave, G., Stillwell, D.: Psychological targeting as an effective approach to digital mass persuasion. Proc. Natl. Acad. Sci. **114**, 201710966 (2017)
McCrae, R.R., Costa, P.T.: Validation of the five-factor model of personality across instruments and observers. J. Pers. Soc. Psychol. **52**(1), 81–90 (1987)
Mcelroy, J., Hendrickson, A., Townsend, A., DeMarie, S.: Dispositional factors in Internet use: personality versus cognitive style. MIS Q. **31**, 809–820 (2007)
Medler-Liraz, H., Yagil, D.: Customer emotion regulation in service interactions: its relationship to employee ingratiation, satisfaction and loyalty intentions. J. Soc. Psychol. **153**(3), 261–278 (2013)
Moon, Y.: Personalization and personality: some effects of customizing message style based on consumer personality. J. Consum. Psychol. **12**(4), 313–325 (2002)
Neff, J.: OMD proves the power of engagement. Advert. Age **78** (2007)
Nguyen, T.H., Waizenegger, L., Techatassanasoontorn, A.A.: Don't neglect the user!—Identifying types of human-chatbot interactions and their associated characteristics. Inf. Syst. Front. **24**, 797–838 (2022)
Nicolescu, L., Tudorache, M.T.: Human-computer interaction in customer service: the experience with AI chatbots—a systematic literature review. Electronics **11**(10), 1579 (2022)
Nyrhinen, J., Uusitalo, O., Frank, L., Wilska, T.A.: How is social capital formed across the digital-physical servicescape? Digit. Bus. **2** (2022)
Pansari, A., Kumar, V.: Customer engagement: the construct, antecedents, and consequences. J. Acad. Mark. Sci. **45**(3), 294–311 (2017)
Parasuraman, A., Berry, L.L., Zeithaml, V.A.: Understanding customer expectations of service. Sloan Manag. Rev. **32**(3), 39 (1991)
Picard, R.W.: Affective computing. MIT, Cambridge, MA (1997)
Piotrowicz, W., Cuthbertson, R.: Introduction to the special issue information technology in retail: toward omnichannel retailing. Int. J. Electron. Commer. **18**, 5–16 (2014)
Prahalad, C.K., Ramaswamy, V.: Co-creating unique value with customers. Strat. Lead. **32**(3) (2004)
Qiu, L., Lin, H., Ramsay, J., Yang, F.: You are what you tweet: personality expression and perception on Twitter. J. Res. Pers. **46**(6), 710–718 (2012)
Radha, S., Prasad, N.: A study on customer orientation as between emotional intelligence and service performance in banks. Int. J. Bus. Manag. Inven. **2**(5), 60–66 (2013)
Robinson, S., Orsingher, C., Alkire, L., De Keyser, A., Giebelhausen, M., Papamichail, K.N., Shams, P., Temerak, M.S.: Frontline encounters of the AI kind: an evolved service encounter framework. J. Bus. Res. **116**, 366–376 (2020)
Ruan, Y., Mezei, J.: When do AI chatbots lead to higher customer satisfaction than human frontline employees in online shopping assistance? Considering product attribute type. J. Retail. Consum. Serv. **68**, 10305 (2022)
Salovey, P., Mayer, J.D.: Emotional intelligence. Imagin. Cogn. Pers. **9**(3), 185–211 (1990)

Schwartz, H.A., Eichstaedt, J.C., Kern, M.L., Dziurzynski, L., Ramones, S.M., Agrawal, M., et al.: Personality, gender, and age in the language of social media: the open-vocabulary approach. PLoS One **8**(9), e73791 (2013)

Scott, M., Zorfas, A., Leemon, D.: The new science of customer emotions: a better way to drive growth and profitability. Harv. Bus. Rev. (2015)

Sedley, R.: Six theses on digital customer engagement in a troubled economy (2008). http://richard-sedley.iuplog.com/default.asp?item=298747

Simon, F., Usunier, J.C.: Cognitive, demographic, and situational determinants of service customer preference for personnel-in-contact over self-service technology. Int. J. Res. Mark. **24**(2), 163–173 (2007)

Simumba, J., Nchito, W.: Customer engagement in tourism and hospitality services in Kasama and Shiwang'andu, Zambia. Afr. J. Hosp. Tour. Leis. 7(2) (2018)

Storbacka, K., Brodie, R.J., Böhmann, T., Maglio, P.P., Nenonen, S.: Actor engagement as a microfoundation for value co-creation. J. Bus. Res. **69**(8), 3008–3017 (2016)

Sun, K., Feng, B., Chen, M., Tao, G.: The impact of core technological capabilities of high-tech industry on sustainable competitive advantage. Sustainability **12**(7) (2020)

Svendsen, G.B., Johnsen, J.K., Sørensen, L.A., Vitterso, J.: Personality and technology acceptance: the influence of personality factors on the core constructs of the technology acceptance model. Behav. Inf. Technol. **32**(4), 323–334 (2013)

Tamir, M.: Why do people regulate their emotions? A taxonomy of motives in emotion regulation. Pers. Soc. Psychol. Rev. **20**(3), 199–222 (2016)

Terzis, V., Moridis, C.N., Economides, A.A.: The effect of emotional feedback on behavioral intention to use computer-based assessment. Comput. Educ. **59**(2), 710–721 (2012)

Thurn, S.: Toward a framework for human-robot interaction. Hum.-Comput. Interact. **19**, 9–24 (2004)

Torkzadeh, S., Zolfagharian, M., Yazdanparast, A., Gremler, D.D.: From customer readiness to customer retention: the mediating role of customer psychological and behavioral engagement. Eur. J. Mark. **56**(7), 1799–1829 (2022)

Tschang, F.T., Almirall, E.: Artificial intelligence as augmenting automation: implications for employment. Acad. Manag. Perspect. **35**(4) (2021)

Verhoef, P., Reinartz, W., Krafft, M.: Customer engagement as a new perspective in customer management. J. Serv. Res. **13**, 247–252 (2010)

Verhoef, P.C., Kannan, P.K., Inman, J.J.: From multi-channel retailing to omni-channel retailing: introduction to the special issue on multi-channel retailing. J. Retail. **91**(2), 174–181 (2015)

Vivek, S.D., Beatty, S.E., Morgan, R.M.: Customer engagement: exploring customer relationships beyond purchase. J Mark Theory Pract **20**(2), 122–146 (2012)

Wang, Q., Woo, H.L., Quek, C.L., Yang, Y., Liu, M.: Using the Facebook group as a learning management system: an exploratory study. Br. J. Edu. Technol. **43**, 428–438 (2012)

Wang, J.N., Du, J., Chiu, Y.-L., Li, J.: Dynamic effects of customer experience levels on durable product satisfaction: price and popularity moderation. Electron. Commer. Res. Appl. **28**, 16–29 (2018)

Wieseke, J., Geigenmüller, A., Kraus, F.: On the role of empathy in customer-employee interactions. J. Serv. Res. **15**(3), 316–331 (2012)

Wirtz, J., Patterson, P.G., Kunz, W.H., Gruber, T., Lu, V.N., Paluch, S., Martins, A.: Brave new world: service robots in the frontline. J. Serv. Manag. **29**(5), 907–931 (2018)

Wirtz, J., Anouk, A., Bloemer, J., Horváth, C., Ramaseshan, B., Van de Klundert, J., Gurhan Canli, Z., Kandampully, J.: Managing brands and customer engagement in online brand communities. J. Serv. Manag. **24**, 223–244 (2013)

Wolfe, K., Kim, H.J.: Emotional intelligence, job satisfaction, and job tenure among hotel managers. J. Hum. Resour. Hosp. Tour. **12**, 175–191 (2013)

Zaid, S., Patwayati, P.: Impact of customer experience and customer engagement on satisfaction and loyalty: a case study in Indonesia. J. Asian Financ. Econ. Bus. **8**(4), 983–992 (2021)

Zheng, P., Wang, Z., Chen, C.-H., Khoo, L.: A survey of smart product-service systems: key aspects, challenges, and future perspectives. Adv. Eng. Inform. **42**, 100973 (2019)
Zmud, R.W.: Perceptions of cognitive styles: acquisition, exhibition, and implications for information system design. J. Manag. **5**(1), 7–20 (1979)

Marisa Faggini is Associate Professor of Macroeconomics, Department of Economics and Statistics, University of Salerno where she teaches courses in macroeconomics, circular economy and laboratory of Europe-an projects. She has a Phd in Economics from University of Naples "Phartenepone". She is author of many publications in national and international journals and her research interests span various domains, including economic policy, complexity and chaos theory in economics, circular economy. Presently, her research endeavors focus on food systems and sustainability on economic systems. From 2008 she is Series Editors with M. Gallegati, A. Kirman and T. Lux of NEW Series, Springer.

Marzia Del Prete holds a Ph.D. in Economics, with a focus on Marketing Management, from the University of Salerno, Italy. Their doctoral thesis, titled ″Emotional Artificial Intelligence: Detecting and Managing Customer Emotions in Automated Customer Care,″ reflects Marzia′s expertise in AI and data-driven strategies. Currently, she is a manager in the AI & Data division of Deloitte Consulting Italy, where she leverages her extensive experience to drive innovative solutions and manage complex projects. Her international experience as a project manager showcases strong leadership skills in integrated EU-project and resource management. With several years of global practical experience in strategic and applied marketing within multinational companies like H3G Italy (CK Hutchison Holdings Lim-ited) and in academia, Marzia offers unique insights and entrepreneurial acumen.

Bruna Bruno is Associate Professor of Economic Policy at the Department of Economics and Statistics of the University of Salerno, Italy, with more than 20 years of teaching and research experience. She has a Phd in Economics from University of Naples Federico II. Author of many publications in indexed journals, she is referee for academic journals. She participated to numerous international scientific conferences as speaker. Her teaching fields include Labor Economics, Public Sector Microeconomics, Public Services Economics and Technological Innovation and Sharing Economy. Research interests cover environmental and innovation eco-nomics, labor economics, education economics, agent-based models, and happiness.

Anna Parziale is Adjunct Professor of Microeconomics and Macroeco-nomics, Department of Business—Management &Innovation Sciences, University of Salerno. After graduating in Economics she holds a doctoral degree in Economic Sciences from the University of Naples Parthenope and a Master in Applied Econometrics from the Higher School of Eco-nomics and Finance in Rome. The training path was also perfected with research experiences abroad including the Universitè catholique de Louvain (Belgium). Author of numerous publications in national and international journals and her research interests span various domains, including economic policy, public economics, chaos theory, economic complex theory, circular economy. Presently, her research endeavors focus on economics of innovation and sustainability on economic systems.

Chapter 9
Phygital Materiality in Service Ecosystems: The Transformative Role of NFTs

Cristina Mele, Tiziana Russo Spena, and Stefano Paolo Russo

Abstract This chapter explores the transformative impact of Non-Fungible Tokens (NFTs) on service ecosystems, emphasising their role in fostering novel forms of resource exchange and value co-creation. It delves into the concept of *phygital* materiality—the seamless integration of physical and digital elements—that NFTs epitomise. Key contributions of the chapter include an analysis of the diverse roles of actors, highlighting virtual agency, new forms of collaboration, and the innovative use of resources to certify ownership and authenticity of items, thereby enabling new modes of exchange. These aspects lay the groundwork for developing new institutional arrangements.

Keywords NFTs · Phygital materiality · Institutional arrangements · Actors · Resources · Value co-creation processes

Introduction

Non-fungible tokens (NFTs) represent a rapidly emerging market with significant opportunities and inherent risks. These digital assets, minted on blockchain platforms like Ethereum, certify ownership and authenticity for unique digital items, ranging from art and music to virtual real estate and collectables. Despite market volatility and ongoing debates about their future, NFTs have gained substantial economic significance: in 2023, the total market performance of NFTs recorded approximately 9.2

C. Mele (✉) · T. Russo Spena · S. P. Russo
University of Naples Federico II, Naples, Italy
e-mail: cristina.mele@unina.it

T. Russo Spena
e-mail: russospe@unina.it

S. P. Russo
e-mail: stefanopaolo.russo@unina.it

D. Sarno et al. (eds.), *Complexity and Emergence in Market Ecosystems*, New Economic Windows, https://doi.org/10.1007/978-3-031-81942-1_9

billion USD in trades (Zuppinger & Laglasse, 2024). This growing interest underscores the need to understand NFTs within the broader context of service ecosystems, highlighting their complexity and innovative properties. Service ecosystems are complex adaptive systems in which multiple actors co-create value through resource integration and service exchange (Vargo & Lusch, 2014).

The service ecosystem framework provides a comprehensive lens for examining how NFTs and other digital innovations drive new forms of value co-creation and service provision. This perspective is especially relevant as it captures the non-linear, emergent properties of interactions, reflecting the mainstream of systems thinking (Polese et al., 2021).

A key aspect of service ecosystems is their focus on materiality, which encompasses the exchange of both tangible and intangible resources within the market. Despite an expanding body of research on service ecosystems, notable gaps remain in understanding the new forms of materiality introduced by digital transformation. The advent of smart technologies has fostered a new form of *phygital* materiality. As Mele and Russo Spena (2024) described, phygital materiality is the seamless integration of physical and digital elements, enabling novel forms of value creation. This concept blurs the boundaries between the tangible and intangible, allowing digital entities to exert material effects and influence within service ecosystems.

This chapter analyses the transformative role of NFTs as new phygital resources that impact actors, resources, and value co-creation processes within service ecosystems. The implications of these aspects lay the groundwork for developing new institutional arrangements. To provide empirical evidence, we employed ethnographic methods, specifically a participant-observation approach conducted via online platforms (Kozinets, 2002). Through digital observation of 25 NFT case studies across various business areas, we gathered data on how NFTs' phygital materiality shapes service ecosystems.

This chapter contributes to the literature on service ecosystems by analysing how NFTs, as novel phygital resources, reshape ecosystem dynamism and structure. Through their unique capacity to bridge the digital and physical realms, NFTs introduce innovative resource exchange and value co-creation modes, prompting the development of new institutional arrangements. These arrangements foster novel interactions and collaborations among actors, redefine resource utilisation, and transform co-creation processes, laying a foundation for evolving norms and practices within service ecosystems.

Service Ecosystems: Complexity and Institutional Arrangements

The conceptualisation of service ecosystems has highlighted the importance of systems in marketing and service literature, rekindling researchers' interest in a systemic perspective on value co-creation (Vargo & Lusch, 2011). This approach is grounded in understanding phenomenological value as emerging through interactions and the application of resources within systems of service-for-service exchange (Vargo & Lusch, 2011). The explanatory power of the service ecosystem metaphor partly lies in its capacity to account for a pluralistic view of business and society (Vargo & Lusch, 2011). Service involves various entities seeking and providing solutions (Lusch et al., 2010), forming a network of interconnected parts that collaborate to co-create value. A service system is initially described as a complex adaptive system, integrating resources and interacting with other service systems via value propositions (Maglio & Spohrer, 2008), which may establish stable relationships within broader service networks. However, this network is not simply the sum of its parts; it includes a higher-order construct formed through interactions. Companies and consumers are complex service systems engaging in activities to achieve desired outcomes (Vargo & Akaka, 2009).

Within this framework, complexity science provides a valuable lens for analysing and understanding service ecosystems, offering a robust conceptualisation of their properties and behaviours. Complexity overcomes traditional dichotomies in systems thinking—such as part versus whole, open versus closed, and individual versus structure (Holbrook, 2003). It is not merely the study of systems with complex causes and effects; as Holbrook (2003) noted, complexity theory examines the rules governing interactions between elements, which generate novel properties at higher levels of the system. In this sense, complexity science advances service ecosystems' systemic and interactional perspective, engaging with non-linear processes and dynamics to emphasise their co-creative and co-adaptive nature (Holbrook, 2003).

Within service ecosystems, social contexts consist of networks of actors and institutional arrangements—sets of institutions embedded at multiple levels of social systems—that shape actions and interactions (Vargo & Lusch, 2011; Vargo et al., 2015). The service ecosystem perspective underscores the co-created nature of value, the dynamic integration of resources, and the role of institutions, or shared "rules of the game," in interconnected systems of service-for-service exchange (Vargo et al., 2015; Vargo & Lusch, 2016). Shared institutional arrangements enable and constrain how resources are integrated and how value is co-created within service ecosystems (Wieland et al., 2016).

The multiple levels of interaction foundational to ecosystems (Akaka et al., 2013) imply a need to zoom in and out between micro, meso, and macro levels, viewing "institutions" as coordination mechanisms (Vargo & Lusch, 2016). Evolving institutions (norms and practices) and broader institutional arrangements act as coordination heuristics that drive value co-creation (Akaka et al., 2013). Institutions evolve as actors continuously co-create them through social interactions involving resource integration and service exchange (Vargo & Lusch, 2017). Institutional maintenance, disruption, and change—or institutionalisation—is a vital mechanism through which the vitality of service ecosystems is produced and reproduced.

The rapid advancement of the digital age has recently influenced service ecosystems and the complex systems of interaction and resource integration that support them. Herterich et al. (2023) explored smart service ecosystems, highlighting socio-technical antecedents such as shared worldviews, structural flexibility, and architectures of participation. Mele and Russo Spena (2024) propose a new tech-based ecology for the service ecosystem, emerging through debates on onlife agency and phygital materiality. This new form of phygital materiality underscores technology's role in facilitating innovative interactions and enhancing value co-creation. Addressing this new materiality enables actors to engage in more dynamic and contextually responsive value co-creation practices, driving the continuous evolution and adaptation of the service ecosystem.

From Digital Materiality to Phygital Materiality

Materiality is often associated with the tangible aspects of artefacts, which are physical objects perceptible through the senses and defined by their boundaries and material substance. In contrast, digital objects—entities accessed via physical devices (e.g., consoles, mobile phones)—lack a permanent physical presence, such as video games or augmented reality artefacts (Magaudda, 2011).

With the advent of the digital age and the proliferation of resources across a continuum from physical to virtual/digital realms, the concept of digital materiality has gained traction. Emerging in architecture, design, media studies, digital anthropology, and digital sociology (Miller & Horst, 2020), digital materiality acknowledges that digital objects and environments, while intangible, possess material properties and impacts. This notion challenges the traditional divide between physical and digital realms by asserting that digital entities exhibit a material dimension through their effects, uses, and the infrastructures supporting them.

Scholars have highlighted various aspects of digital materiality across diverse contexts, including design and social media (Davis & Beck, 2023). Kirschenbaum (2008) differentiates between forensic and formal types of digital materiality: forensic materiality involves the physical evidence of production, distribution, reception, and preservation, which is observable through visible or instrument-detectable "traces" In contrast, formal materiality pertains to digital media's architecture and symbolic structures, including individual software designs, embedded data standards, or metadata encoding (Shep, 2015). Furthermore, some scholars argue that materiality should not be solely defined by the presence of tangible resources (Leonardi, 2010). A digital artefact, though lacking physical attributes, can be deemed material because it exists beyond a conceptual domain, imposing constraints and affordances akin to physical artefacts. This perspective suggests that materiality is not an inherent property of artefacts but emerges from the interactions between artefacts and the individuals who produce and use them (Leonardi, 2010). From this standpoint, physical properties are tangible features and qualities that enable or constrain actions (Mele & Russo Spena, 2024). Thus, digital materiality is not a finished product but a process of becoming, achieved through the performativity of artefacts that equip people with capabilities to meet goals (Leonardi, 2010).

In summary, rather than offering a singular definition, digital materiality can be understood as a process arising from a series of interactions rather than a static quality of matter (Pink & Fors, 2017). Adopting this concept reveals key aspects: infrastructure, data as material, user interaction, and environmental impact. Digital activities are supported by physical infrastructure—servers, data centres, and networks—which require material resources and energy for storage, transmission, and processing. Additionally, users engage with digital interfaces through physical devices like smartphones, tablets, and computers, underscoring the material interaction with digital content. The production, use, and disposal of these digital devices and infrastructures have significant environmental consequences, highlighting the materiality of digital technology (Berkhout & Hertin, 2004).

Once digital materiality is established, the concept of phygital further emphasises the integration of material and digital. Phygital materiality refers to the convergence of digital technologies, social presence, and physical elements (Mele et al., 2021; Mele & Russo Spena, 2024). Despite its varied applications—ranging from objects and spaces to shopping experiences and customer journeys (Mele et al., 2023)—phygital materiality merges the physical and digital worlds to create experiences where these boundaries are blurred (Mele, Di Bernardo et al., 2024). This concept has increasing relevance in retail, marketing, and service design, where integrating physical and digital components can enhance user experience and engagement. Key characteristics of phygital materiality include interaction, interactivity, personalisation, and immersive experiences: modern technologies seamlessly blend physical and digital elements (e.g., augmented reality overlays digital information on the physical world); user interaction is heightened by the merging of physical and digital interfaces (e.g., retail store touchscreens providing product information or virtual trials); digital data enhances the personalisation of physical experiences (e.g., smart home devices adjusting lighting and temperature based on user preferences); and immersive environments engage multiple senses through technologies like virtual reality, simulating physical spaces or offering interactive installations in galleries.

In conclusion, digital and phygital materiality are evolving concepts that reflect the dynamic nature of interactions and value co-creation in an increasingly technological society. While digital materiality focuses on the tangible aspects of digital entities, phygital materiality emphasises the seamless integration of the physical and digital realms to enhance co-creation. An illustrative example of this new materiality is the rise of NFTs.

Non-Fungible Token

In recent decades, the concept of Web3 has gained considerable traction (Rudman & Bruwer, 2016). NFTs are created (or "minted") on a blockchain, such as Ethereum (Wang et al., 2021), and represent units of data that certify a digital asset as unique and non-interchangeable, offering a distinct digital certificate of ownership (Nadini et al., 2021). NFTs are blockchain-enabled assets that signify ownership of art, songs, newspaper articles, and viral video clips (Chalmers et al., 2022).

Much of the literature on NFTs has focused on their nature and technical features (Wang et al., 2021). Without delving into excessive technical details, we can begin to understand the nature of NFTs by outlining their key characteristics: uniqueness, authenticity, reliability, non-interchangeability, scarcity, and collectibility (Chandra, 2022). Each NFT, minted on the blockchain, represents a unique object and has no identical counterpart in terms of attributes, such as form, content, name, and blockchain address (Sestino et al., 2022). Furthermore, the blockchain technology underlying NFTs ensures that each asset is irreversible and unchangeable once minted, thus guaranteeing its authenticity (Guadamuz, 2021). This assurance of authenticity and uniqueness enables each NFT to be easily bought and resold on the secondary market, generating additional revenue for its creators (Kräussl & Tugnetti, 2024). NFTs function as cryptographic tokens, but unlike cryptocurrencies, they are not interchangeable—each NFT includes a "smart code" that distinguishes it from others (Bao & Roubaud, 2022). NFTs are also typically produced in limited quantities or collections, often centred around a specific theme (Chohan, 2021), making them effective tools for collecting various types of virtual objects (Valeonti et al., 2021; Russo et al., 2024).

The technical features of NFTs make them suitable for various applications or contexts, categorised into six primary types: Art, Collectibles, Games, Metaverse, Utility, and Other (Nadini et al., 2021). The first category includes NFTs as digital artworks, such as images, videos, or GIFs (Russo et al., 2024); collectables comprise NFTs across various fields (e.g., fashion, museums, music) of interest to collectors (Valeonti et al., 2021; Guadamuz, 2021); NFT games apply within gaming contexts or ecosystems (Wang et al., 2021; Muthe et al., 2020); metaverse NFTs represent elements of virtual worlds (Hwang, 2023); utility NFTs serve specific purposes (e.g., access to events or financial services) (Kong & Lin, 2021); and the "Other" category covers NFTs in small collections not included in the other categories (Nadini et al., 2021).

Regardless of their type or application, the benefits and opportunities generated by NFTs can be grouped into three main themes: transparency and security, new business opportunities, and the development of the metaverse (Wang et al., 2021; Bhujel & Rahulamathavan, 2022). NFTs provide significant transparency and security through blockchain technology, their fundamental infrastructure (Bao & Roubaud, 2022). Every transaction involving an NFT's creation, sale, or exchange is permanently recorded on the blockchain. Companies can use NFTs to enhance the transparency and security of product data, offer exclusive access to special events or products, and promote their brands with unique digital collectables (Sestino et al., 2022; Wang et al., 2021; Chandra, 2022). Lastly, NFTs serve as foundational resources in the metaverse, enabling a digital asset market within a Web3 environment where ownership and distribution rules remain flexible (Mele et al., 2024a, b).

In conclusion, NFTs are phygital resources that can transform or create new service ecosystems. However, the debate within service research remains in its infancy, and most studies focus on technical aspects. An essential contribution is to explore how NFTs impact service ecosystems and the value co-creation process.

Research Process

We employed netnography, a participant-observational study conducted via online resources (Kozinets, 2002). This qualitative research approach is well-suited for deep understanding and detailed description of complex phenomena. In emerging research areas, interpretative data collection methods are particularly valuable (Eisenhardt, 1989), as they facilitate gathering thick data that captures a given scenario's narratives, implications, and nuances. Through digital observations, we gathered implicit and often subtle data points (Fusch & Ness, 2015).

Our research process was divided into two main phases. First, data collection was conducted from March 2023 to March 2024. We searched online for information-rich case studies (Piekkari et al., 2010). The objective was to generate sufficiently detailed material to illuminate the patterns, concepts, and dynamics of the phenomena under study (Gummesson, 2005). Following Stake's (1995) guidelines, we selected cases to maximise insights while considering time and resource constraints. We reviewed company websites, NFT marketplaces, and other online resources to identify companies or projects launching new NFT initiatives.

Additionally, blog posts and social media content were utilised to capture the social construction and sensemaking associated with the phenomenon. We also supplemented web data with information from journal articles, case reports, and other relevant materials (Bowen, 2009). A snowball sampling allowed us to expand our case collection until we reached a saturation point, where no new data emerged (Lincoln & Guba, 1985). This approach resulted in a judgmental sample of 25 NFT solutions (Table 9.1) across diverse businesses, thereby adding robustness and depth to the framework (Järvensivu & Törnroos, 2010).

Second, we conducted a qualitative content analysis (Bryman, 2008; Heinonen & Medberg, 2018), involving four steps: open, axial, reflective, and selective coding (Scott & Howell, 2008). In the open coding phase, we used thematic analysis to elevate raw data to a conceptual level (Gummesson, 2017), with two researchers independently reviewing the material to identify key categories. Axial coding was then applied to link and group these categories, identifying the central phenomenon and its patterns. We expanded the codes during this phase and re-examined the dataset (Scott & Howell, 2008). Next, we created reflective coding matrices structured as a relational hierarchy to identify the core category, subcategories, and their interrelationships (Strauss & Corbin, 1998). By analysing multiple themes and providing detailed descriptions, we enhanced the explanatory power of the empirical study and developed a conceptual framework (Halinen & Törnroos, 2005). To capture the interrelationships among categories, researchers engaged in selective coding through an interpretive process, identifying conditions and dimensions (Scott &

Table 9.1 Sample

Company/Initiative	Industry	NFTs solutions
Alfa Romeo	Automotive	NFTs assigned to vehicles to provide transparent and reliable information to potential buyers when the vehicle is resold
Bored Ape Yacht Club	Web3	The Sandbox platform offers the possibility of transforming the NFTs of Bored Ape Yacht Club members into avatars
Burberry	Fashion	Collaboration with gaming company Mythical Games aims at launching a collection of exclusive NFTs
CloudKeys	Web3	Partnership aimed at creating exclusive NFTs for luxury travel destinations
Coca Cola	Food and beverage	NFT collectibles
Dolce & Gabbana	Fashion	Fashion collection, which consists of clothing items that live both in their physical and digital forms through Art NFTs that represent them
Everyday: The First 5000 Days	Art	Digital artwork supported by NFT
Ferrari	Automotive	NFT collectibles
Food conglomerates	Food	Food-associated NFTs provide consumers with full transparency on the origin and route of their products
Freed's Bakery	Food	NFT collection depicting cakes that can also be purchased in physical form
Gucci	Fashion	A unique collection of customised, consumer-facing NFTs representing the essence of the brand created through a partnership with a digital artist
Illuvium Lab	Gaming	A game platform designed to be interoperable with other games and blockchain projects by enabling players to buy, sell and trade NFT as rewards and other items on the market for profit
Mattel	Toy	A collection of rare toy cars with physical die-cast cars and NFT cards featuring them
McDonald's	Food	Limited-edition digital collectibles featuring the iconic sandwich
NBA Top Shot	Sport	Basketball-themed NFT community
Nike	Sport	NFT collections and the creation of a personal platform for NFTs
Nomo Soho Hotel	Hospitality	Regular NFT booking offer
Saysoon	Art	Certification of crypto and physical works via NFT
Sorare	Sport	Opportunity to experience football by collecting, trading, buying, selling and competing with NFT player cards
Stargraph	Web3	Digital autographs are exchangeable via NFT certificates

(continued)

Table 9.1 (continued)

Company/Initiative	Industry	NFTs solutions
The Sandbox	Web3	NFT avatar that allows users the ability to navigate the Sandbox metaverse while enjoying unique benefits or rewards
Travala	Hospitality	A travel rewards programme based on collectable NFTs gives traveller-users access to various benefits beyond simple booking
Warner Music Group	Music	Partnership with OpenSea platform aimed at giving musical artists the possibility to build new Web3 communities and create new forms of connection and creativity powered by NFTs
Yellowheart	Ticketing	Sale of tickets in the form of NFT to reduce the risk of fraud, NFT and provide additional benefits
Yuga Labs	Web3	A collection on the Ethereum blockchain with cartoon-style profile images

Howell, 2008). Finally, we discussed the findings collaboratively to reach a consensus (Gummesson & Lee, 2005).

NFTs in Service Ecosystems

NFTs play a transformative role in service ecosystems by introducing new materiality that integrates components such as blockchain networks, digital wallets, marketplaces, creator platforms, and community forums. These components create a novel bridge between the physical and digital worlds, enabling the seamless creation, distribution, and exchange of resources. This integration fosters a dynamic environment of phygital ownership and innovation.

Incorporating NFTs into service ecosystems impacts actors, resources, and value co-creation processes. We explore these aspects in detail below.

Actors

Within service ecosystems, various actors benefit from new opportunities, experiences, and modes of engagement by creating, exchanging, and selling NFTs. The technical properties of NFTs foster novel interactions among actors and encourage the development of new skills and capabilities. Companies, users, platforms, and other stakeholders leveraging NFT potential can choose how to position themselves and define their roles within NFT-based service ecosystems. For example, when used as avatars, NFTs enable new forms of virtual agency and user interaction. In online or virtual spaces, users can select their digital appearance and behavioural traits through avatars—digital representations of their identities within a digital environment. Avatars need not be humanoid; they can take various forms, from animals to artistic images. Recently, avatars have increasingly been created as NFTs, fully exploiting the potential of this technology. The "Bored Ape Yacht Club" by Yuga Labs has become one of the most popular NFT projects, featuring cartoon-style profile images built on the Ethereum blockchain. More advanced NFT avatars are 3D characters that users can manoeuvre within virtual worlds. For instance, the Sandbox platform launched the "Solo Leveling: Unlimited Avatar Collection," which allows NFT avatar owners to navigate the Sandbox metaverse while enjoying special perks or rewards, such as access to exclusive events with other Web3 users, varying based on the rarity and characteristics of the NFT avatar.

NFTs also provide opportunities to enhance collaboration within service ecosystems, enabling creators, platforms, and consumers to access and exchange resources that promote collective growth and innovation. Integrating artistic, technical, and business skills is essential to channel NFT technology into valuable digital assets. For example, Warner Music Group (WMG), a global music and entertainment company, partnered with the OpenSea platform to help WMG artists build new Web3 communities and introduce existing fan communities to novel NFT-powered connections and creative opportunities. NBA Top Shot is another example of a platform where users can own and trade digital collectables of NBA and WNBA highlights, known as Moment NFTs, while engaging with the community to discuss basketball NFTs, ask questions, and share insights. In a similar spirit of collaboration, Gucci partnered with a digital artist to launch the "10KTF Gucci Grail" project, a collection of customised NFTs that reflect the brand's identity. Likewise, Burberry's collaboration with gaming company Mythical Games resulted in a series of NFTs, granting owners exclusive access to virtual spaces and real-world rewards.

Additionally, CloudKeys, in partnership with Hospitality Services PC, a company active in the holiday rental industry, has pioneered the use of blockchain and NFTs to offer select travellers the chance to explore luxury travel destinations, merging digital and physical experiences in novel ways.

Resources

NFTs are considered vital resources in the Web3 realm (along with cryptocurrencies) because they facilitate a secure marketplace for digital assets (e.g., images, videos, songs, and tweets) within digital or phygital worlds, where ownership and distribution rules can be challenging to define. NFTs can certify the ownership and authenticity of a wide range of items, acting as an information keeper. For instance, in virtual marketplaces and metaverses, NFTs ensure that digital items like art, collectables, and virtual real estate are unique and cannot be duplicated, thereby maintaining their value and desirability. Stargraph, for example, developed one of the first digital autograph platforms, allowing artists to sign digital autographs in limited editions and distribute them to select fans through NFT certificates.

NFTs not only secure ownership and uniqueness of digital assets; they can also store, verify, and share information about the provenance, authenticity, and past use of physical assets, equipping market participants with a resource aimed at transparency and efficiency. These aspects are critical in ecosystems where the uniqueness and origin of items are essential, such as virtual marketplaces and the metaverse. For example, Alfa Romeo's Tonale model uses blockchain technology to assign a unique NFT to each vehicle, enabling comprehensive tracking of the vehicle lifecycle and providing transparent, reliable information for potential buyers. Similarly, in the food sector, many companies are adopting NFTs to offer consumers full transparency on the origin and journey of their products, helping to ensure the authenticity and quality of food purchases.

The connection between the digital and physical dimensions enables the creation of complex resources known as phygital NFTs, which integrate physical and digital elements. Unlike traditional digital assets or simple digital representations of physical items, phygital NFTs combine both dimensions into a unified entity. This concept reimagines the design and exchange of composite goods, enhancing user experiences by merging tangible and intangible assets. For instance, Freed's Bakery created a collection of NFTs featuring hand-painted cakes on the OpenSea marketplace, with each depicted cake available for purchase as a physical item for local pickup or delivery. Dolce & Gabbana's "Genesis" collection similarly consists of clothing items in both physical and digital forms through Art NFTs representing each piece. Likewise, Mattel has launched a collection of rare toy cars, each accompanied by both a physical die-cast car and an NFT card representing it.

Whether purely digital or phygital, NFTs enable new forms of exchange due to their inherent characteristics of uniqueness, ownership verification, and ease of transfer within the blockchain ecosystem. This capability supports new economic models, such as fractional ownership and resale royalties, which were difficult to implement before the advent of blockchain technology. NFTs are highly valuable in contexts like digital artwork, virtual real estate, and digital collectables, where value is often derived from scarcity, the reputation of the creator, and community interest. NFTs also allow digital content creators to monetise their work directly, bypassing traditional intermediaries. Notable examples include the digital art pieces "Everydays: The First 5000 Days" and "Forever Rose," which demonstrate how digital artists have successfully marketed and sold purely digital works that can garner significant interest from investors and collectors.

In the gaming industry, digital items acquired by players can be converted into NFTs, granting players ownership of in-game resources or even entire game elements. These assets can then be traded or exchanged, transforming game tokens into valuable and potentially profitable digital assets. Numerous brands across diverse industries, including Gucci, Coca-Cola, Ferrari, and McDonald's, have also launched collectible NFTs (NFT drops) to promote their vision and offer consumers new ways to engage with the brand. Sorare, a startup, has created an innovative virtual sports game based on blockchain and trading cards, appealing to sports enthusiasts by enabling users to collect, trade, buy, sell, and compete with NFT player cards, providing an interactive football experience for fans worldwide.

Value Co-Creation Processes

The potential of NFT technology extends beyond creating new resources; it provides a foundation for creators, users, and platforms to develop innovative methods of interaction and value co-creation. New mechanisms structure and guide these interactions within the NFT ecosystem.

NFTs offer a standardised method for representing and exchanging digital assets across various platforms and ecosystems, preserving their properties and characteristics to enable cross-platform interoperability and system integration. This feature is particularly relevant in the metaverse, allowing digital objects to be transportable and usable across multiple virtual environments. Seamless integration of virtual microworlds—such as Roblox, Decentraland, and The Sandbox—is seen as a prerequisite for building a true metaverse: a unified virtual world that parallels the real one. A significant attempt at interoperability was made by The Sandbox platform, which allows Bored Ape Yacht Club members to transform their NFTs into avatars. Users can connect to The Sandbox by linking their wallets, enabling a voxel version of their NFTs to appear.

Furthermore, gaming represents one of the primary domains where NFT interoperability reaches its full potential, giving players greater flexibility by enabling them to integrate resources across different games. For example, Illuvium by Illuvium Lab is a sci-fi virtual world that combines open-world exploration with a collection of NFT creatures. The platform is designed for interoperability and allows players to buy, sell, and trade NFTs as rewards and in-game items.

Integrating NFTs into service ecosystems can foster innovation in how services are designed and delivered. NFTs can tokenise access to services, experiences, or membership benefits, offering unique and personalised service options. Tokenised services provide both utilitarian benefits (e.g., reliability, transparency, security) and hedonic benefits (e.g., brand membership, community sense, exclusivity). Yellowheart, a company specialising in selling NFT tickets, reduces fraud risk and ensures that ticket issuers receive payment on every resale. NFT tickets function as exclusive access passes, reimagining what it means to be part of a community with privileges and rewards. Similarly, the Nomo Soho hotel became the first in the U.S. to offer regular reservations through NFTs, with benefits such as late checkout and free breakfast, creating a membership club experience for guests. Travala, one of the world's leading blockchain-based travel booking platforms, has created a travel rewards program featuring collectable NFTs, providing users access to benefits beyond standard bookings.

Finally, NFTs open new revenue models for businesses and individuals within service ecosystems. Companies can monetise digital content, intellectual property, and virtual goods by selling and trading NFTs, a particularly lucrative art, music, and entertainment model where digital creations can be tokenised and sold directly to consumers. Additionally, creators can earn royalties on secondary sales of their NFTs, securing ongoing revenue from their work. For example, Nike earned over $180 million in 2022 through its NFT collections and the creation of a proprietary NFT platform. Digital artworks and other NFTs created by individuals or digital art groups continue to command high prices on major NFT marketplaces. Saysoon, an innovative company in the art sector, certifies both crypto and physical artworks via NFTs, bridging the gap between traditional and crypto art markets. This system ensures that artists benefit from the appreciation in value over time, as they receive a commission every time their works are resold, not just upon the initial purchase.

Discussion

This chapter explores the complex dynamics within service ecosystems transformed by the introduction of NFTs, with particular emphasis on *phygital* materiality—a unique fusion of physical and digital properties enabled by NFTs. We examine how NFTs drive new institutional arrangements within these ecosystems, impacting three core aspects: actors, resources, and value co-creation processes. Through this perspective, we reveal how NFTs enhance the role of actors, influencing interactions and resource access. These assets seamlessly blend tangible elements, such as physical artworks or merchandise, with digital enhancements like blockchain authentication and augmented reality features (Dwivedi et al., 2022).

Our findings indicate that phygital materiality, which integrates physical and digital elements, fosters new engagement and value co-creation forms. This integration significantly impacts service ecosystems by redefining interactions among actors, resources, and processes (Mele & Russo Spena, 2024). Our framework (Fig. 9.1) highlights the critical elements essential for understanding the transformative potential of NFTs.

First, NFTs have revolutionised the concept of digital agency and interaction, particularly within the Web3 ecosystem. Users can adopt NFT avatars in virtual spaces, enhancing their digital agency and enabling novel forms of interaction. The ability to customise digital appearances and behavioural traits fosters more personalised and dynamic engagement, strengthening users' sense of individuality and presence in virtual spaces. NFTs also facilitate innovative collaborations, enabling new interactions and capabilities. Within these ecosystems, creators, platforms, and consumers engage in cooperative ventures that drive collective growth and innovation. By leveraging NFTs, actors can co-create content, share intellectual property, and participate in mutually beneficial partnerships.

NFTs serve as critical resources within the Web3 domain, certifying ownership and authenticity of both pure digital and phygital assets, which is essential in environments where the uniqueness and provenance of goods are paramount. NFT technology has led to tokens with varying levels of integration with the physical world. Depending on their use and specific attributes at minting, NFTs span

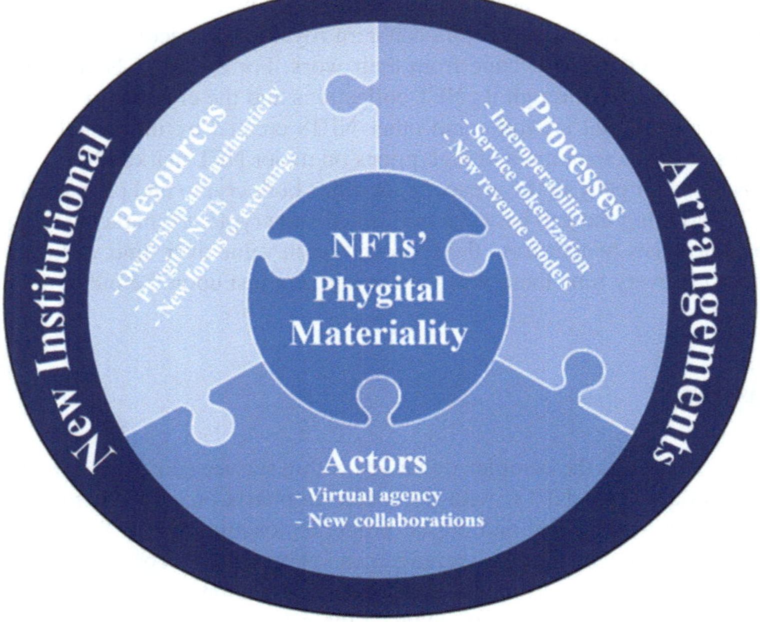

Fig. 9.1 NFT-based service ecosystems. *Source* Authors' elaboration

a spectrum from digital to physical applications. In virtual marketplaces and the metaverse, the capacity to verify the authenticity and provenance of digital items fosters trust and reliability, facilitating secure, seamless transactions. NFTs create new forms of resource exchange by enabling the buying, selling, and trading of assets on blockchain-based platforms, establishing a novel economic layer where goods acquire new intrinsic value. These economies allow users to create value from digital activities, blending virtual and physical economic systems.

Thirdly, NFTs drive innovative value co-creation processes by standardising the representation and exchange of assets across different platforms and ecosystems. Evolving standards and practices guide actors' interactions and resource integration (Vargo & Lusch, 2014). Interoperability is particularly relevant in the metaverse, where the portability and usability of digital items across various virtual environments are essential. NFTs support a cohesive and interconnected digital ecosystem by ensuring digital assets retain value and functionality across platforms. NFTs can also tokenise access to services, offering personalised and exclusive benefits that drive user engagement and create a sense of exclusivity and value. This innovative use of NFTs enables new revenue models by monetising digital content, intellectual property, and virtual goods. Artists, musicians, and companies can tokenise their work, generating direct sales revenue and royalties from secondary market transactions.

In conclusion, NFTs play a transformative role in service ecosystems, giving rise to new institutional arrangements by enhancing digital ownership, fostering collaboration, enabling new forms of exchange, and driving innovative value co-creation. As phygital resources, NFTs reshape actors' roles, process dynamics, and the nature of resources within service ecosystems. This new phygital materiality supports the development of smart, adaptive, and responsive service ecosystems where physical and digital elements coexist and interact fluidly, promoting opportunities for value co-creation (Mele & Russo Spena, 2024).

Limitations and Further Research

Despite NFTs' transformative potential in service ecosystems, several limitations hinder their full realisation, particularly regarding actors, resources, and processes. These limitations also suggest avenues for further research (see Table 9.2).

First, different actors—including companies, users, and platforms—face challenges within NFT-based ecosystems. The usability of NFT platforms remains a significant barrier to engagement and user experience, as many consumers are largely

Table 9.2 Further research

Focus	Research question
Actors' engagement and experience	How can the user experience of NFT platforms be improved to enhance accessibility for non-technical users? What economic models can be developed to reduce volatility and promote long-term engagement in the NFT-based ecosystem market? How can legal frameworks be designed to provide clear guidelines on NFT ownership and intellectual property rights?
Efficient use of resources	What greener technologies (i.e. blockchain) and design practices can be developed to reduce the environmental impact of NFTs? How can NFT interoperability be improved to enable seamless use across multiple platforms? What technological solutions can enhance blockchain platforms' scalability and transaction speeds supporting NFTs?
Facilitating value co-creation processes	How can standardised protocols for representing and exchanging NFTs across different platforms be developed to facilitate cohesive ecosystems? What advanced security measures can be implemented to protect against smart contract vulnerabilities and fraud in NFT transactions? How can processes for tokenising services and benefits be innovated to create personalised offerings for users?

unaware of NFTs and their numerous applications. To enhance accessibility, NFT platforms should simplify interfaces with intuitive designs, guided onboarding, and educational resources, ensuring mobile-friendly access and easy navigation. Such improvements can create a smoother and more accessible experience for newcomers. Additionally, the high volatility of the NFT market creates uncertainty, deterring long-term investments and participation. Research should focus on developing more stable and predictable economic models to reduce market volatility and encourage sustained engagement from diverse actors. Legal ambiguities surrounding NFT ownership and intellectual property rights further complicate actors' ability to engage confidently. Comprehensive studies are needed to clarify the regulatory landscape, providing clear guidelines on ownership and intellectual property rights.

Second, although NFTs are essential resources within Web3 ecosystems, several factors limit their effectiveness. Due to the energy-intensive nature of blockchain transactions, the environmental impact of NFTs poses significant sustainability concerns. Research should explore the development of carbon–neutral technologies (e.g., greener blockchains) and design practices to mitigate these impacts, such as transitioning to more energy-efficient consensus mechanisms and implementing carbon offset strategies. Furthermore, limited interoperability across different platforms restricts the usability and value of NFTs. The current technological infrastructure, including blockchain scalability and transaction speeds, remains inadequate for large-scale adoption. Research should aim to improve the interoperability and infrastructure supporting NFTs, enabling seamless transfer and use across multiple platforms and virtual environments, thereby increasing their value and usability.

Third, value co-creation and exchange processes within NFT ecosystems face several challenges. The absence of standardised protocols for representing and exchanging NFTs across platforms impedes seamless interactions and value co-creation. Future research should focus on developing these standardised protocols to facilitate more cohesive and integrated ecosystems. Security concerns, such as smart contract vulnerabilities and the risk of fraud, threaten the integrity of NFT transactions. Strengthening transaction security is essential and can be achieved through robust smart contract designs and advanced authentication methods. Prioritising these measures will help protect users from fraud and unauthorised access, fostering greater trust in the NFT ecosystem.

Lastly, while promising, tokenising services and benefits processes are still underdeveloped, limiting their potential for creating personalised and exclusive offerings that could also impact physical spaces, such as on-site shopping experiences. Research should explore innovative ways to tokenise services and benefits, enhancing user engagement and value creation through unique and personalised offerings.

References

Akaka, M.A., Vargo, S.L., Lusch, R.F.: The complexity of context: a service ecosystems approach for international marketing. J. Int. Mark. **21**(4), 1–20 (2013)

Bao, H., Roubaud, D.: Non-fungible token: a systematic review and research agenda. J. Risk Financ. Manag. **15**(5), 215 (2022)
Berkhout, F., Hertin, J.: De-materialising and re-materialising: digital technologies and the environment. Futures **36**(8), 903–920 (2004)
Bhujel, S., Rahulamathavan, Y.: A survey: security, transparency, and scalability issues of NFTs and its marketplaces. Sensors **22**(22), 8833 (2022)
Bowen, G.A.: Document analysis as a qualitative research method. Qual. Res. J. **9**(2), 27–40 (2009)
Bryman, A.: Of methods and methodology. Qual. Res. Organ. & Manag.: Interna-Tional J. **3**(2), 159–168 (2008)
Chalmers, D., Fisch, C., Matthews, R., Quinn, W., Recker, J.: Beyond the bubble: Will NFTs and digital proof of ownership empower creative industry entrepreneurs? J. Bus. Ventur. Insights **17**, e00309 (2022)
Chandra, Y.: Non-fungible token-enabled entrepreneurship: a conceptual framework. J. Bus. Ventur. Insights **18**, e00323 (2022)
Chohan, U.W.: Non-fungible tokens: Blockchains, scarcity, and value. In: Chohan, U.W. (ed.) Non-Fungible Tokens, pp. 1–11. Routledge, London (2021)
Davis, D.J., Beck, T.E.: How social media disrupts institutions: exploring the intersection of online disinformation, digital materiality and field-level change. Inf. Organ. **33**(4), 100488 (2023)
Dwivedi, Y.K., Hughes, L., Baabdullah, A.M., Ribeiro-Navarrete, S., Giannakis, M., Al-Debei, M.M., Wamba, S.F., et al.: Metaverse beyond the hype: multidisciplinary perspectives on emerging challenges, opportunities, and agenda for research, practice and policy. Int. J. Inf. Manag. **66**, 102542 (2022)
Eisenhardt, K.M.: Building theories from case study research. Acad. Manag. Rev. **14**(4), 532–550 (1989)
Fusch, P.I., Ness, L.R.: Are we there yet? Data saturation in qualitative research. Qual. Rep. **20**, 1408–1416 (2015)
Guadamuz, A.: The treachery of images: non-fungible tokens and copyright. J. Intellect. Prop. Law Pract. **16**(12), 1367–1385 (2021)
Gummesson, E.: From relationship marketing to total relationship marketing and beyond. J. Serv. Mark. **31**(1), 16–19 (2017)
Gummesson, E., Lee, N.: Qualitative research in marketing: road-map for a wilderness of complexity and unpredictability. Eur. J. Mark. **39**(3/4), 309–327 (2005)
Halinen, A., Törnroos, J.Å.: Using case methods in the study of contemporary business networks. J. Bus. Res. **58**(9), 1285–1297 (2005)
Heinonen, K., Medberg, G.: Netnography as a tool for understanding customers: implications for service research and practice. J. Serv. Mark. **32**(6), 657–679 (2018)
Herterich, M.M., Dremel, C., Wulf, J., vom Brocke, J.: The emergence of smart service ecosystems—The role of socio-technical antecedents and affordances. Inf. Syst. J. **33**(3), 524–566 (2023)
Holbrook, M.B.: Adventures in complexity: an essay on dynamic open complex adaptive systems, butterfly effects, self-organising order, coevolution, the ecological perspective, fitness landscapes, market spaces, emergent beauty at the edge of chaos, and all that jazz. Acad. Mark. Sci. Rev. **6**(1), 1–184 (2003)
Hwang, Y.: When makers meet the metaverse: effects of creating NFT metaverse exhibition in maker education. Comput. Educ. **194**, 104693 (2023)
Järvensivu, T., Törnroos, J.Å.: Case study research with moderate constructionism: conceptualisation and practical illustration. Ind. Mark. Manage. **39**(1), 100–108 (2010)
Kirschenbaum, M.G.: New media and the forensic imagination. The MIT Press, London (2008)
Kong, D.R., Lin, T.C.: Alternative investments in the Fintech era: the risk and return of non-fungible token (NFT). (2021). https://doi.org/10.2139/ssrn.3914085
Kozinets, R.V.: The field behind the screen: using netnography for marketing research in online communities. J. Mark. Res. **39**(1), 61–72 (2002)

Kräussl, R., Tugnetti, A.: Non-fungible tokens (NFTs): a review of pricing determinants, applications and opportunities. Journal of Economic Surveys **38**(2), 555–574 (2024)
Leonardi, P.M.: Digital materiality? How artifacts without matter, matter. First Monday **15**(6) (2010)
Lincoln, Y.S., Guba, E.G.: Naturalistic inquiry. Sage Publications Inc., Beverly Hills (1985)
Lusch, R.F., Vargo, S.L., Tanniru, M.: Service, value networks and learning. J. Acad. Mark. Sci. **38**, 19–31 (2010)
Magaudda, P.: When materiality 'bites back': digital music consumption practices in the age of dematerialisation. J. Consum. Cult. **11**(1), 15–36 (2011)
Maglio, P.P., Spohrer, J.: Fundamentals of service science. J. Acad. Mark. Sci. **36**, 18–20 (2008)
Mele, C., Russo-Spena, T.: Agencement of onlife and phygital: smart tech-enabled value co-creation practices. J. Serv. Manag. (First Line) (2024). https://doi.org/10.1108/JOSM-03-2023-0113
Mele, C., Russo-Spena, T., Tregua, M., Amitrano, C.C.: The millennial customer journey: a phygital mapping of emotional, behavioural, and social experiences. J. Consum. Mark. **38**(4), 420–433 (2021)
Mele, C., Russo Spena, T.R., Marzullo, M., Di Bernardo, I.: The phygital transformation: a systematic review and a research agenda. Ital. J. Mark. **2023**(3), 323–349 (2023)
Mele, C., Russo Spena, T., Russo, S.P.: Experimenting on the metaverse to foster innovation. Span. J. Mark.-ESIC (First Online) (2024a). https://doi.org/10.1108/SJME-05-2023-0117
Mele, C., Di Bernardo, I., Ranieri, A., Russo Spena, T.: Phygital customer journey: a practice-based approach. J. Cetacean Res. Manag. **27**(3), 388–412 (2024b)
Miller, D., Horst, H.A.: The digital and the human: A prospectus for digital anthropology. In: Miller, D., Horst, H.A. (eds.) Digital anthropology, pp. 3–35. Routledge, London (2020)
Muthe, K.B., Sharma, K., Sri, K.E.N.: A blockchain-based decentralised computing and NFT infrastructure for game networks. In: 2020, the Second International Conference On Blockchain Computing And Applications (BCCA) (pp. 73–77). IEEE (2020)
Nadini, M., Alessandretti, L., Di Giacinto, F., Martino, M., Aiello, L.M., Baronchelli, A.: Mapping the NFT revolution: market trends, trade networks, and visual features. Sci. Rep. **11**(1), 20902 (2021)
Piekkari, R., Plakoyiannaki, E., Welch, C.: 'Good' case research in industrial marketing: insights from research practice. Ind. Mark. Manage. **39**(1), 109–117 (2010)
Pink, S., Fors, V.: Being in a mediated world: self-tracking and the mind-body–environment. Cult. Geogr. **24**(3), 375–388 (2017)
Polese, F., Payne, A., Frow, P., Sarno, D., Nenonen, S.: Emergence and phase transitions in service ecosystems. J. Bus. Res. **127**, 25–34 (2021)
Rudman, R., Bruwer, R.: Defining Web 3.0: opportunities and challenges. Electron. Libr. **34**(1), 132–154 (2016)
Russo, S.P., Mele, C., Russo Spena, T.: Innovative value propositions in the fashion metaverse. J. Glob. Fash. Market. **15**(1), 39–61 (2024)
Scott, K.W., Howell, D.: Clarifying analysis and interpretation in grounded theory: using a conditional relationship guide and reflective coding matrix. Int J Qual Methods **7**(2), 1–15 (2008)
Sestino, A., Guido, G., Peluso, A.M.: The interplay of consumer innovativeness and status consumption orientation when buying NFT-based fashion products. In: Sestino, A., Guido, G., Peluso, A.M. (eds.) Non-Fungible Tokens (NFTs) examining the impact on consumers and marketing strategies, pp. 63–75. Springer International Publishing, New York (2022)
Shep, S.J.: Digital materiality. In: Schreibman, S., Siemens, R., Unsworth, J. (eds.) A NEW COMPANION TO DIGITAL Humanities, (pp. 322–330). Chichester: John Wiley & Sons (2015)
Stake, R.: Case study research. Sage Publications Inc., Thousand Oaks (1995)
Strauss, A., Corbin, J.: Basics of qualitative research: techniques and procedures for developing grounded theory. Sage Publications Inc., Newbury Park (1998)
Valeonti, F., Bikakis, A., Terras, M., Speed, C., Hudson-Smith, A., Chalkias, K.: Crypto collectibles, museum funding and OpenGLAM: challenges, opportunities and the potential of Non-Fungible Tokens (NFTs). Appl. Sci. **11**(21), 9931 (2021)

Vargo, S.L., Akaka, M.A.: Service-dominant logic as a foundation for service science: clarifications. Serv. Sci. **1**(1), 32–41 (2009)

Vargo, S.L., Lusch, R.F.: It's all B2B… and beyond: toward a systems perspective of the market. Ind. Mark. Manage. **40**(2), 181–187 (2011)

Vargo, S.L., Lusch, R.F.: Service-dominant logic: What it is, what it is not, what it might be. In: Vargo, S.L., Lusch, R.F. (eds.) The service-dominant logic of marketing, pp. 43–56. Routledge, London (2014)

Vargo, S.L., Lusch, R.F.: Institutions and axioms: an extension and update of service-dominant logic. J. Acad. Mark. Sci. **44**, 5–23 (2016)

Vargo, S.L., Lusch, R.F.: Service-dominant logic 2025. Int. J. Res. Mark. **34**(1), 46–67 (2017)

Vargo, S.L., Wieland, H., Akaka, M.A.: Innovation through institutionalisation: a service ecosystems perspective. Ind. Mark. Manage. **44**, 63–72 (2015)

Wang, Q., Li, R., Wang, Q., Chen, S.: Non-fungible token (NFT): overview, evaluation, opportunities and challenges. *arXiv preprint* arXiv:2105.07447 (2021)

Wieland, H., Koskela-Huotari, K., Vargo, S.L.: Extending actor participation in value creation: an institutional view. J. Strateg. Mark. **24**(3–4), 210–226 (2016)

Zuppinger, G., Laglasse, M: 2023 Annual market report (2024). Access at https://nft18.com/reports/2023-nft-annual-report/

Cristina Mele Full Professor in the Department of Economics, Management, and Institutions at the University of Naples "Federico II." An expert in service innovation, smart technologies, markets, service ecosystems, and value co-creation, she has authored approximately 300 publications, including contributions to the Journal of The Academy of Marketing Science, Journal of Business Re-search, Industrial Marketing Management, Journal of Service Man-agement, Journal of Retailing and Consumer Services, Journal of Business Market Management, Service Science, International Journal of Quality and Reliability Management.

Tiziana Russo Spena Full professor of Business Management and Coordinator of Ph.D. in Management at the University of Naples Federico II. She has been involved as local unit coordinator of many scientific projects on in-novation and technology transfer. Her main areas of interest are ser-vice innovation and digital transformation. She has co-authored over 150 refereed articles, six books on innovation topics and international Journals, including the Industrial Marketing and Management Journal of Business Ethics, Journal of Service Management, Journal of Business Research, and Journal of Service Theory and Practices.

Stefano Paolo Russo Ph.D. candidate at the University of Naples Federico II. His main areas of interest are service innovation, digital transformation and social robotics.

Chapter 10
Navigating Digital Transformation in Cultural Organizations: Balancing Authenticity, Tradition, and Innovation

Saba Samadilashkariani and Jon Engström

Abstract As cultural service organizations increasingly integrate digital technologies, they face critical challenges in maintaining their authenticity—a concept deeply rooted in their traditional practices. This study explores the negotiation of authenticity within cultural organizations as a function of value co-creation. In line with the interactionist perspective on authenticity, we approach authenticity as the outcome of ongoing interactive negotiations. Viewing authenticity through the negotiation lens offers a fresh perspective on the context-specific dynamics of value co-creation. The study builds on a qualitative study of the Royal Swedish Opera, during the transition from de-institutionalization phase to re-institutionalization phase of digitalization. We conducted semi-structured interviews with 25 employees in various key positions, from administrative staff to artistic leadership. Our findings reveal that authenticity is negotiated among various actors in three dimensions concerning Historical Continuity, Artistic Commitment, and Social Commitment. The study sheds light on how negotiations are essential to the emergence of new solutions and novel comprehensions of authenticity and value. Using Hegel's dialectics, this chapter demonstrates how a synthesis emerges as a resolution to the paradox of honoring tradition while pursuing innovation. Importantly, our study highlights the need to understand digitalization in the light of forces present at the different levels of the ecosystem, and the necessity to view digitalization as the means to renegotiate organizational and societal goals.

Keywords Authenticity · Interactive negotiations · Digital transformation · Value co-creation · Phase transformation

S. Samadilashkariani (✉) · J. Engström
Stockholm Business School, Stockholm University, Stockholm, Sweden
e-mail: saba.samadilashkariani@sbs.su.se

J. Engström
e-mail: jon.engstrom@sbs.su.se

D. Sarno et al. (eds.), *Complexity and Emergence in Market Ecosystems*, New Economic Windows, https://doi.org/10.1007/978-3-031-81942-1_10

Introduction

As service organizations and their offerings are undergoing digitalization, organizations and the actors in them still have reservations about this transition. Fear of losing the traditions and authenticity that are deeply ingrained in the service offering is an example of the challenges of digitalization (Keiningham et al., 2019). The risks associated with innovation are particularly evident in cultural services, where adherence to tradition, symbolic meanings, and authenticity is crucial for establishing market legitimacy, and the introduction of digitalization presents a potential threat to these deeply rooted elements (Lampel et al., 2000). Cultural organizations typically prioritize symbols, traditions, and authentic productions over market-driven strategies (Bourdieu, 1990). For cultural services, the digitalization that is now occurring represents a significant phase transition (cf. Polese et al., 2021). Centuries-long traditions of producing and consuming culture are going from being fully physical to digital, or a hybrid between physical and digital.

Examples of digitalization in cultural services range from the digital display of images by museums (Bertacchini and Morando, 2011; Wang and Lin, 2018), to the virtual tours and online galleries offered by institutions such as the Louvre Museum and The Opera di Santa Maria del Fiore in Florence (Evrard and Krebs, 2018; Fissi et al., 2021), and virtual concerts with CGI productions, such as ABBA Voyage. Digitalization as illustrated in these examples, offers significant opportunities to enhance cultural services' outreach, design customer-centered strategies, and foster deeper community engagement (De Bernardi et al., 2019; Fraser et al., 2017; Hodgkiss et al., 2019). However, digitalization also creates tension with the social, cultural, and artistic missions that these institutions uphold. These tensions include the conflict between artistic and economic values, as well as the tensions between adherence to tradition and the embrace of novelty and differentiation (cf. Lampel et al., 2000). These tensions are further complicated by the specific nature of the organization; for instance, museums face tensions between maintaining control and opening up access (Bertacchini and Morando, 2011), while performing arts organizations contend with concerns regarding the dilution of the essence of the art form and its immediacy in the digital age (King, 2018; Morris, 2010; Trubnikova and Tsagareyshvili, 2021). Furthermore, the heritage of these organizations is crucial in serving as a source of cultural creativity and artistic expression (Eyerman and Jamison, 1998). It encompasses embedded cultural myths and narratives (Dacin and Dacin, 2007), which are at risk of being eroded as the medium transforms to digital (Hartmann and Ostberg, 2013).

Digital transformation represents a significant phase transition where old structures are destabilized by internal and external disturbances, and new resources, value, institutional arrangements, and practices, which are necessary for attaining stability in the new state, emerge (Polese et al., 2021). Digitalization, thus, not only challenges old ways of working, but also notions of value and authenticity. The existing body of research on how cultural organizations navigate such phase transitions is notably limited. In particular, there is a lack of exploration into the interactive negotiation processes through which new notions of authenticity are co-created and tradition and innovation are reconciled. This co-creation of authenticity is a compelling and underexplored area in service research, that has implications for the services, the arts, and our collective cultural heritage.

In this chapter, we are interested in how organizations in the cultural sector address the dilemmas they face during digitalization. To examine the process of re-institutionalization during the phase transition, we studied the digitalization of the Royal Swedish Opera. The study builds on qualitative interviews with 25 employees across a broad spectrum of roles, ranging from administrative staff to artistic directors. We adopt the interactive negotiation approach to authenticity, to analyze the negotiations that take place as the organization balances novelty and tradition. The interactive negotiation approach addresses authenticity as "*not something that exists, does not exist, or should exist as an inherent or made property of a person, artifact, object, or place, but as part of an interactive, culturally informed process of negotiation*" (Beerends and Aydin, 2021, p. 1672). Drawing on Hegel's dialectic, this study investigates the interaction between contrasting perspectives: the thesis of upholding traditions and the antithesis of embracing innovation. The study contributes to an increased understanding of the dimensions in which authenticity negotiation takes place in the context of a cultural service organization, and of the dynamics in these negotiations. By focusing on these aspects, we elaborate on the interactive authenticity negotiations among diverse actors, demonstrating how self-organization (cf. Polese et al., 2021; Lusch and Vargo, 2014; MacLean and MacIntosh, 2003) is a process of aligning differing perspectives and integrating resources, which is highly influenced by internal conceptions and the conditions in the broader service ecosystem. The study responds to recent calls for deeper exploration into the influence of actor interactions on service ecosystem emergence and phase transition (Polese et al., 2021). In line with the foundational view that value co-creation involves dynamic and interactive processes among actors (Vargo and Lusch, 2004), we view authenticity as a co-created dimension that contributes to how value is created in service activities, aligning with scholars who explore diverse facets of the concept (e.g., Kelleher et al., 2019; Beirão et al., 2017).

Understanding the impacts of digitalization on cultural service organizations particularly, carries significant societal implications, especially in terms of expanding access to cultural services and preserving cultural traditions. Viewing authenticity through the lens of negotiation offers a fresh perspective on the context-specific dynamics of value co-creation. The integration of diverse technological tools not only transforms the nature of service activities (cf. Edvardsson and Tronvoll, 2022) but also leads to the emergence of new notions of value, interactively shaped in the organization. Managerially, the study is relevant particularly for service organizations seeking to leverage their traditions amidst digital transformation. It offers universal lessons for various service organizations to reinterpret and innovate upon their traditions.

Theoretical Background

Phase Transition and Emergence in Service Ecosystem

Phase transition refers to significant shifts within a system prompted by external disturbances and internal interactions, leading to the de-institutionalization of existing structures, followed by re-institutionalization with new institutional arrangements and value, and eventually a new stable state (Polese et al., 2021). The notion of phase transition helps explain how systems adapt their institutions to maintain stability while encountering disturbance and change. This concept is particularly relevant in the context of digital transformation in service ecosystems, where traditional practices are disrupted and new configurations emerge (Orlikowski and Scott, 2021). In the context of cultural organizations, digital transformation represents a significant phase transition, which not only challenges ways of working, but also traditional notions of authenticity.

Institutions refer to the "regulative, normative, and cultural-cognitive elements that, together with associated activities and resources, provide stability and meaning to social life" (Scott, 2014, p. 56) which are essential facilitators of value co-creation among systems (Vargo et al., 2015; Vargo and Lusch, 2016; Vink et al., 2021). Hence, the process of re-institutionalization, where new norms and practices become established, is crucial in understanding how ecosystems adapt to changes such as digitalization. Emergence is the process by which new properties, behaviors, or patterns arise from the interactions of the components of a system (Polese et al 2021). For cultural organizations undergoing digital transformation, emergence involves the reinterpretation of traditions in a digital context, leading to innovative ways of engaging with audiences and creating value (Russo-Spena and Mele, 2012). Digital technologies enable these organizations to reach wider audiences and offer new forms of interaction, thereby expanding their service ecosystem (Hanna and Rowley, 2011). Still, it requires negotiating the balance between cultural heritage and innovation.

The Co-Creation of Authenticity

Descriptions of Authenticity

The concept of authenticity is often associated with the genuineness and truth that characterizes an individual, object, or artifact, demonstrating its alignment with originality and historical accuracy (Grayson and Martinec, 2004). Several attributes are associated with the concept of authenticity, such as truth, genuineness, realness, sincerity, and trustworthiness, which are essential for brands to cultivate an image of integrity and responsibility (Beverland and Farrelly, 2010; Södergren, 2021; Morhart et al., 2015). Brands also reinforce their authenticity by demonstrating social commitment, production methods, heritage, and a commitment to quality while minimizing commercial motives (Fritz et al., 2017; Beverland, 2006). A theme that particularly often emerges in discussions of authenticity relates to the commitment to preserving traditions and heritage. Morhart et al. (2013) refer to this dimension of authenticity as continuity, highlighting that the brand's stability and historical presence, indicate its likelihood of persistence in the future. In their study on Music and Social Movements, Eyerman and Jamison (1998) similarly argue that traditions play a vital role in legitimizing music by distinguishing authentic music production from commercial production. While traditions foster authenticity and market legitimacy, excessive reliance on traditions can lead to organizational inertia and dogmatism (March, 1991). Therefore, it is crucial to continuously reconstruct and interpret traditions (Giddens et al., 1994), to enable the formation of innovative practices that align with valued rituals (Cancellieri et al., 2022). Similarly, other authenticity-attributions, such as the method of production, quality commitment, and social commitment, need to be continuously revisited and reinterpreted.

Paradigmatic Perspectives on the Creation of Authenticity and Value

Authenticity has been viewed through different paradigmatic perspectives in academic research. Two significant perspectives of authenticity can be discerned: the essentialist perspective, which considers authenticity as an inherent quality of the object, artifact, or place, and the constructivist perspective, which asserts that authenticity is socially constructed by observers' subjective perspectives (Beverland, 2006; Beverland and Farrelly, 2010; Fritz et al., 2017; Grayson and Martinec, 2004). From the essentialist perspective, authenticity is derived from the object or person itself, such as an authentic Picasso painting (Beverland, 2006; Grayson and Martinec, 2004). This view posits that authenticity is an intrinsic attribute that can be objectively measured and is not subject to individual perceptions (Morhart et al., 2013). In this stream, the authenticity of brands is judged from objective sources of information, such as origin and craftsmanship. Thus, brands aim to signal objectively authentic attributes in their products and services, for instance, through connection to the past and representation of traditions (Beverland, 2005; Jones et al., 2005).

On the other hand, the constructivist perspective posits that authenticity is a social and cultural construct (Arnould and Price, 2000; Hartmann and Ostberg, 2013). It's an attribute shaped by observers' subjective experiences and cultural contexts. In this stream, authenticity is not an objective property but a perception that is interpreted differently by individuals, influenced by their personal and cultural frames (Arnould and Price, 2000). For example, in the realm of craft production, Hartmann and Ostberg (2013) argue that authenticity is not just a matter of using traditional techniques or materials, but also how these methods and products are perceived and valued within the community. Thus, the constructivist perspective acknowledges that different groups may have varying criteria for what counts as authentic. The discourse has shifted from understanding authenticity as an intrinsic phenomenon tied to an object, person, or product, to focusing on authenticity as co-created through interactive practices or authenticating acts (Peterson, 2005). In this view, authenticity is not only perceived but actively co-constructed through interactions and social practices (Beerends and Aydin, 2021; Hartmann and Ostberg, 2013). Along the same lines, an *interactionist perspective* suggests that authenticity is continuously *negotiated* among various stakeholders, framing it as a culturally informed interactive process of negotiations (Beerends and Aydin, 2021).

The academic discourse on authenticity and how it emerges parallels the discourse in marketing on value and the creation of value in significant ways. Marketing scholars have approached the concept of value from multiple paradigmatic perspectives: from the positivist paradigm on value as an objective and measurable reality, to the interpretivist paradigm which assumes value as subjective and interpreted by individuals, the discourse has evolved to increasingly viewing value as socially constructed, recognizing inherent qualities as well as the subjective experiences and social contexts that shape perceptions of value (Zeithaml et al. 2020). Value in this view is multidimensional; through what dimensions value should be understood, and how each dimension should be understood, is subject to negotiations in the social constructivist's view (Zeithaml et al., 2020). The discourse on value and value creation in the service marketing literature is especially closely aligned with the social constructivist perspective, perceiving value as co-created in social or institutional arrangements, while simultaneously drawing from objectivists and interpretivism, such in the definition of value as subjective and phenomenological (Brodie et al., 2019). The recent focus on networks in service research has reinforced the social-constructivist tendencies in service research and added a zoomed-out view of the value co-creation from individual to collective and social, offering additional structural details (Vargo and Lusch, 2016). Beirãoet al. (2017) showed the value co-creation from micro to meso- and macro-level, proposing that with a systemic view, value outcomes at different levels of aggregation can be deduced.

In our view, authenticity is part of the broader umbrella of value and is co-created in tandem with other value co-creating activities, and just as value, it has outcomes at individual, organizational, and societal levels. When we study authenticity in the context of digitalizing the opera, we anticipate that questions of authenticity will emerge in a large array of value co-creation instances. What constitutes authenticity and value more broadly is an especially important question in the context of a phase transition such as digitalization, which introduces significant change and turbulence.

Negotiations of Authenticity

The process of negotiating authenticity is also referred to as *authentication*: the continuous process of claiming, achieving, or bestowing authenticity among various actors (cf. Arnould and Price, 2000; Beverland and Farrelly, 2010; Peterson, 2005). For instance, in the context of Country Music, Peterson (2005) found that the assessment of authenticity is determined by a sequence of actors, including artists, producers, record companies, vocalists, backup musicians, radio programmers, listeners, and fans. Each actor has the power to accept or reject the authenticity of the music, thus impacting the authentication process. The act of authentication in this context arises from the fusion of individual life experiences embedded within a narrative of self-development (Arnould and Price, 2000), suggesting that each person's authentication journey is fundamentally grounded in their unique life paths and experiences. Similarly, Beerends and Aydin (2021) propose that authenticity is better comprehended as an outcome of negotiation between various social actors. They elucidate: "*Instead of authenticity-attributions confirmed by experts, which remain unchallenged, authenticity is negotiated by various social agents acting within everyday-life situations. From this perspective, (im)material artifacts are not as such authentic or inauthentic—or essentially authentic or authentically constructed—but rather understood as outcomes of negotiations between various stakeholders*" (Beerends and Aydin, 2021, p.166).

The negotiations can be understood as a power struggle, where the social and cultural frameworks, backgrounds, positions, and expectations of the individuals involved in negotiations, along with their subjective interpretations of authenticity, all play a significant role (Arnould and Price, 2000; Beverland and Ferelly; 2010). Actors leverage their respective forms of capital—cultural, social, and economic (c.f Bourdieu, 1984)—to exert influence in negotiations. In this view, authenticity serves as a mechanism of social control that mirrors power relations (Beerends and Aydin, 2021).

Negotiations can also be viewed through the lens of dialectics, as a process of resolving contradictions and opposing forces to arrive at a more advanced or synthesized position. The dialectic method, proposed in 1807, involves reconciling opposing elements to reach a higher understanding. In Hegelian terms, the *thesis* represents the established traditions and practices that define an organization's identity. These traditions, while integral, must evolve over time to stay relevant. The *antithesis* introduces new challenges, such as digital technologies, which disrupt these traditional practices and force organizations to confront and reassess their foundational values (Maybee, 2020). The opposing elements, thesis, and antithesis, consist of contradictory yet interrelated elements (Smith and Lewis, 2011). As such, they do not necessarily present a binary choice of either-or but instead entail seemingly paradoxical, yet credible, scenarios (O'Driscoll, 2008). The *synthesis* emerges as the resolution of this conflict, integrating traditional values with innovative advancements. Importantly, this synthesis is not a static endpoint but becomes a new thesis, leading to further dialectical cycles of adaptation and negotiation (Maybee, 2020).

Methodology

This study builds on an in-depth case of the Royal Swedish Opera focusing on its initiative to digitally transform while striving to maintain its authentic service offerings. The choice of a case study was based on the advantages of in-depth insights and the empirical phenomenon it provided (Campbell, 1975; Dubois and Gadde,

Table 10.1 Interviewee titles and category

Steering group Legal Director, Production Director, Technical Director, Communications & Sales Director	Artistic Board Artistic Director at Royal Ballet, Director of Orchestra, Director of Young-at-opera, Deputy Director of Young-at-opera
Project group Producer A, Web & Content Manager, Corporate Lawyer, Dramaturg, Broadcasting, Video Producer	External artistic directors Choreographer
Customer team Key Account Manager, CRM and Analysis, Digital Manager, Sales Manager, System and Development Manager	Enablers IT Manager, Producer B, Producer C, Head of Planning, Support and Service

2002; Yin, 2018), with the Royal Opera being a suitable case for studying questions of digitalization and authenticity, as it is both a forerunner in the digitalization of cultural service organizations and a national cultural institution where authenticity is of utmost importance.

Data

We collected the data through semi-structured, in-depth interviews with 25 employees at the Royal Swedish Opera. A purposive sampling approach was utilized to increase the depth of understanding of the phenomenon (Campbell, 1975). With assistance from the CEO and the head of communications and sales, we selected the most relevant candidates, to ensure a thorough understanding of the processes involved in the negotiations at the opera. We collected nearly 27 h (1574 min) of interviews, with individual sessions ranging from 29 to 72 min in duration (see Table 10.1). All interviews were transcribed verbatim, resulting in 361 pages of text, which was subsequently stored and coded in the Nvivo software. We complemented interview data with secondary data sources, including annual reports, internal organizational documents, the company website, and news articles about the opera.

Analysis

We followed a thematic analysis (Gioia et al., 2013), to examine the perspectives of various participants and the unanticipated insights (Braun and Clarke, 2006). After familiarizing ourselves with the depth and breadth of the content (Nowell et al., 2017) we identified terms and quotes in the interview transcripts and secondary data that seemed relevant to the phenomenon of study, yielding 169 first-order codes. We assigned codes to insights that were considered relevant to the concepts of negotiations as well as challenges with digitalization, as experienced by the interviewees. We also assigned codes to insights that related to the notion of authenticity. Here, guided by our theoretical pre-understanding, we related divergent terms used by the interviewees, such as originality, realness, aesthetic value, traditional value, meaningful, and meaningless, to the concept of authenticity.

We analyzed the initial 169 codes to establish first-order categories, aiming to stay faithful to the interviewee's terms and the keywords in the secondary data (Gioia et al., 2013). This resulted in 47 first-order categories. Next, we reviewed these materials to find similar and paradoxical evidence (Eisenhardt and Graebner, 2007). During this process, we consistently consulted the existing literature and made comparisons with the identified codes and coding structure. Drawing upon the emergent data, pre-existing literature, and theoretical concepts, we employed a cyclic process to establish a tandem between data and theory, adopting an abductive approach (Alvesson and Kärreman, 2007). As a result, seven second-order themes were developed, which aggregated the first-order categories in a way that each theme, delineated a unique tension. Finally, at a higher level of abstraction, we classified the seven second-order themes into three aggregated dimensions (See Fig. 10.1 for a representation of the data structure).

Findings

Background: Towards the Tipping Point of Digitalization

History, the Opera as a Reflection of Its Time

For the last 250 years, the institution of the Royal Swedish Opera has been a cultural scene that, on the one hand, has been a symbol of continuity and conservatism, and on the other hand always reflected the times and the powers in charge. It has shifted dialectically between polarities of openness to the world and nationalism, between conservatism and experimentation, and between addressing the few or the many.

The opera was founded during the Enlightenment era in 1773 by King Gustav III. While the king had a personal love for the opera, the establishment of an opera was also a way to form a national identity, raise the country's reputation, and the creation of an arena for social interactions with the elites. The king fired the French opera group that had previously performed opera in the city, in order to establish an authentically Swedish ensemble, and commanded the construction of an opera house. Ten years after the opening, the king was killed at a masquerade ball at the opera, murdered by a member of the nobility, disgruntled by reforms increasing the king's power over the nobility. The event is the basis for Verdi's opera "Un Ballo in Maschera". While the king died, the opera lived on, and transformed from being a royal opera for the royals and the nobility, to address a broader audience among the

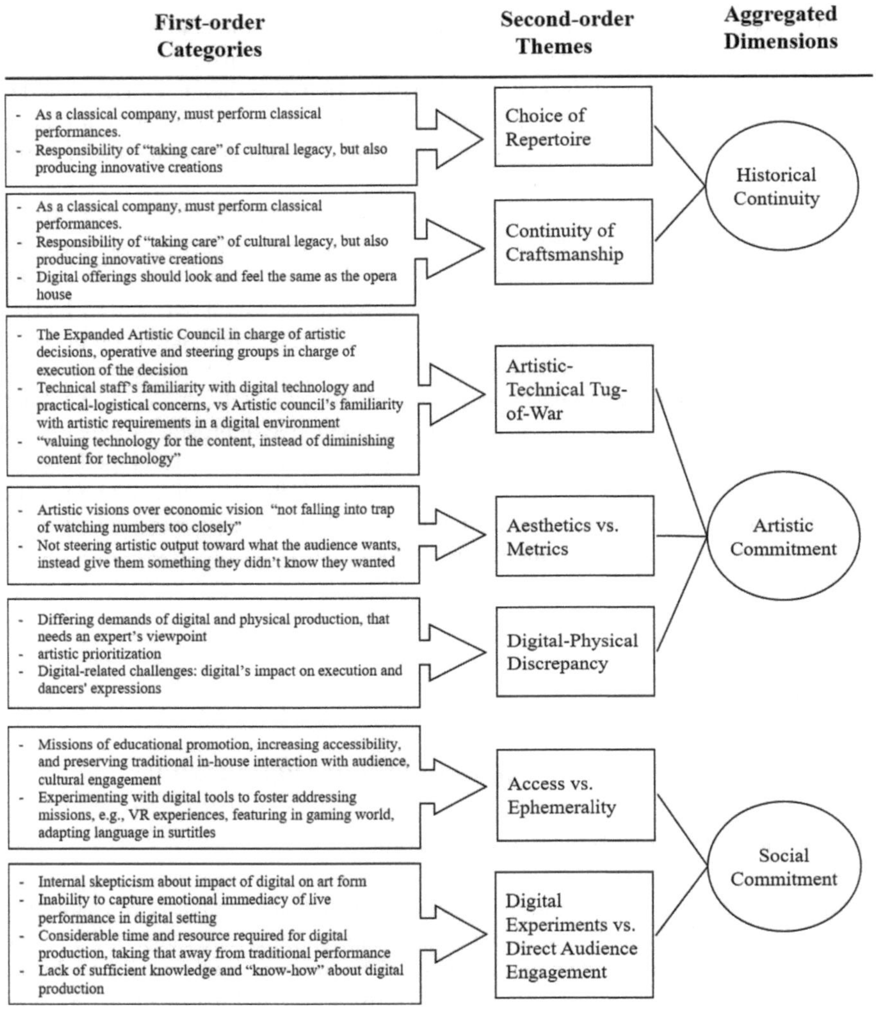

Fig. 10.1 Study's data structure. *Source* Authors' elaboration

bourgeoisie, and access was granted by buying tickets instead of invitations. In the middle of the nineteenth century, the repertoires often included "Grand Operas", by French and Italian composers, that would treat political or other emotion-evoking conflicts and were typically sung in the original language.

In 1898 a new opera house was inaugurated. At the inauguration, only Swedish composer's works were played, reflecting the increasing national romantic tendencies of that era. During the world wars, the opera was a place where anxieties could be expressed through performances. In the following years, the country was increasingly democratized and modernized, which was reflected in the opera's offering which now included more experimental and contemporary works. During the 1970s, cultural establishments enjoyed increased financial support from the government, much of it used to make culture accessible to all through outreach programs in line with social democratic programs, including a touring opera, staging performances in other cities (initiatives that were mostly stopped due to financial constraints).

The Opera in the 2000s

The latter parts of the century, until today, have been signified by a more balanced approach of tradition and novelty, but with a new emphasis on sustainability and inclusion, and a desire that the opera mirrors the increasingly ethnically diverse population. As the emphasis on outreach was a key goal, digitalization of the opera was viewed as a way to meet the goals of reaching out, while remaining within budget. Digitalized performances were shown in movie theatres across the country, with the purpose of reaching a broader audience. Internationally, digitalization efforts were brewing: The Met in New York started to live stream live performances in 2006, followed by other similar initiatives globally, including the European Union's initiative Operavision, a collaboration of European Operas that allows the public to enjoy opera recordings through their web browsers. The tipping point, at which the need for a more concerted effort for digitalization was reached, happened in 2018, when the government mandated the opera to make performances accessible to all the country, through digital broadcasts. The mandate also stressed the dual aim of maintaining traditions and the craft, while also developing the art form, push for international exchange, and focusing on a diverse audience.

In parallel with the increased focus on digital offerings to the end customer, digitalization of the back end also increased, supporting both administrative tasks and the production of physical and digital performances. On the front end, digitalization facilitated the provision of streaming options, the archiving of performances digitally, the improvement of the online ticketing system, and the creation of innovative productions, including a Virtual Reality ballet performance launched in Spring 2021. In March 2020, the opera introduced its digital platform, Operanplay, featuring a handpicked selection of exclusive performances, available online and on demand. On the back end, digital tools were employed to optimize internal communication through the deployment of advanced planning software and an intranet system. These new systems enabled seamless connectivity among all employees, facilitating efficient information sharing and automated updates. Additionally, digital tools were integrated into the production process. These tools included a 3D visualization studio that allowed the stage to be modeled in three dimensions for remote external collaborations, upper and under-stage machinery installations that facilitated stage movement during performances, and video installations that added visual digital presentations to on-stage performances.

Negotiations of Authenticity

Our analysis revealed conflicting perspectives and ideas regarding employees' perceived authenticity in the context of digitalization. We identified the following aggregated dimensions: Historical Continuity, Artistic Commitment, and Social Commitment. Within each of these dimensions, multiple paradoxes are negotiated, ultimately contributing to the co-creation of a perception of authenticity. Following the sequence of the aggregated dimensions in the data structure (Fig. 10.1), we detail the identified paradoxes and the negotiation process within each dimension.

Historical Continuity

The first dimension of negotiations of authenticity pertains to negotiations of historical continuity. This refers to the negotiations regarding how to maintain a connection to the historical legacy of the opera during digitalization, both in terms of the final product and in the methods of production. At the opera, and as mandated by the government, there is a dual, and perhaps contradictory, objective of preserving traditions while at the same time embracing innovation, addressing new audiences, and digitalizing.

Choice of Repertoire. The necessity to negotiate historical continuity predates the era of digitalization. In the choice of repertoire, there has long been a tension between traditions and novelty, where the forces for novelty both stem from the political leadership that wants to enforce political ideas to be reflected in the repertoire, and from the artistic leadership, where artistic directors and producers may want to express their creativity through the repertoire. Nevertheless, as the national stage for opera and ballet, the artistic council was compelled to uphold a repertoire that encompasses classical and historically significant works. As noted by the sales manager: "*We can't have five modern ballets during one season because we will not get enough audience ... We are a classical company. We must dance Swan Lake, Nutcracker, and everything like that*".

In the annual reports, the balancing of continuity and innovation was a central theme each year, where the CEO of the opera commented on how the opera's dual mandates –reflecting both political directives and artistic visions—were realized through the selected repertoire. However, due to the unprecedented impact of the COVID-19 pandemic, a more pronounced shift towards digitalizing a larger part of the repertoire became a natural course of action. The CEO commented on this transformative shift in their remarks, in the annual report: "*On March 13, 2020, the Opera's last performance was played for the public. In 24 h, the Opera had to adjust for the next performance, which was streamed live for the digital channel Operanplay, for the first time. Smaller pilot projects had previously been done for three weeks from the Guldfoajén and now it concerned a large Wagner production from the main stage. The Opera thus entered the digital performing arts, which has since been the playground for the loyal, new, and young audience with a unique number of visitors, 864,910 compared to 230,314 the year before*".

Following the pandemic, digital productions presented a challenge to the notion of historical continuity. Although these digital productions themselves did not necessitate a shift in the direction of the repertoire, maintaining these was costly, and in this balance, physical production was prioritized. Thus, despite the success of digital productions and the large number of viewers, the initial focus on digital productions was tempered. In 2023, the CEO noted a decline in digital performances and the number of viewers. The more radical approach of focusing on broad audiences that was initially adopted was adjusted to a more moderate approach, where the digital offerings were viewed as complementary rather than central (cf. Cancellieri et al., 2022).

Continuity of Craftmanship. The issue of historical continuity in the production of performances, particularly in relation to original staging techniques, was another topic of negotiation. The opera had a legal and cultural mandate from the government to preserve traditional skills, but simultaneously a modernization of various elements of production took place. An example of negotiation concerning craftsmanship concerned the use of projectors, rather than paintings, to create the background decor. One interviewee noted the preference for traditional painted walls among certain members of the production team, stating: "*We have projections that were a hit a couple of years ago, everyone wanted projections in the background because you can do a lot with a projector… it's practical but it doesn't have the same feeling. … the painters, really hate the projections because they like doing things old school way. We also have that in the government's letter of assignment, we have to keep the artistic knowledge live. … keep the old handcraft*".

The extent to which traditional methods were valued and applied was in part following generational lines, where younger cohorts displayed a preference for contemporary methodologies, such as building 3D models, over conventional physical models: "*[there are] more and more younger people coming into this, also in the technical parts, scenographers and lighting designers and things like that. But we still have some of the older ones as well, they are coming to the opera with a physical model of how exactly the stage should look like*".

Thus, a seemingly gradual transition towards digital methods of production, as a function of the old guard retiring, with younger generations taking over with a different view on what constitutes an authentic opera production.

With digitalization becoming an inevitable necessity, the opera found alternative ways to promote its historical continuity. For instance, through the special production for the 250th anniversary of the house, the opera promoted its heritage and roots. Furthermore, with the launch of the operanplay, the design was done in a manner that reflects the opera's physical house. Although the acoustics, live staging, and immediate social experience could not be replicated on the digital platform, the visual representation aimed to mirror the opera house. As the communication manager stated: "*We thought this platform [operanplay] should be our new stage so the look and feel of it should be very much as the Royal Swedish Opera is*".

Overall, interviews have demonstrated a clear emphasis on the promotion of heritage and pedigree, two authenticity attributes that are well discussed in the literature (Beverland and Ferelly 2010; Beverland, 2006). These statements echo the importance of continuity in authenticity as emphasized by Morhart et al., (2015), who argue that a brand's ability to remain stable over time indicates its likelihood of enduring in the future. However, an unwavering commitment to historical accuracy can sometimes limit opportunities for creative modernization and undermine cultural relevance. This underscores the importance of engaging in negotiations.

Artistic Commitment

The second dimension of negotiations of authenticity is negotiations of artistic commitment. It refers to negotiations of how to manage commitment to artistic values, such as technical skills of performers, artistic interpretation, choreography and staging, narrative coherence, ensemble performance, and aesthetic impact, in relation to opportunities and challenges to digitalization. These negotiations typically involved the artistic council (consisting of artistic directors) and the operative and steering groups (comprising producers, marketing managers, and technical managers).

Artistic-Technical Tug-of-War. The operational structure of the opera involved a complex framework. Starting from the idea stage, the decision-making process was initiated at a high level, by the Expanded Artistic Council. The artistic council was tasked with setting the strategic direction and artistic vision of the opera, for instance, which performances were included in the repertoire, and how much funding should be allocated to various productions. Once the artistic council made its decisions, feedback was provided by the operative and steering groups, composed of representatives from the production, marketing, and technical departments, who were in charge of planning and executing the artistic choices. During this process, initial artistic choices were continually refined through feedback, to align with logistical and resource considerations. For instance, if the artistic council envisioned a ground production but decided on a limited budget, it was up to the operative group to assess whether this vision could be achieved. As one producer explained: *"If the Artistic Council says we don't need too many resources [for a project], but [it should] look like this and that, then it's the technical department that needs to say we can't have it both. Either, it's this one very expensive looking [piece] …or we will have consequences artistically*".

However, with the advancement of digitalization in the opera house, the need for decentralization of artistic decision-making arose. Certain specialized roles, such as the technical staff, had more proficiency in digital technology, making their input for artistic decisions, such as possibilities of innovative directions to follow, valuable. Conflicts, however, often emerged between their visions and those of the artistic council. For instance, a technical staff highlighted how digital tools, such as LED screens could streamline logistical aspects when changing scenes in a performance, while also enhancing the artistic value by adding a new dimension to the performance.

Conversely, an artistic director expressed skepticism, arguing that incorporating LED screens might reduce the performance to an advertisement-like experience. He stated: *"First of all, technically I think we can do it, it's important to note. And second, what would be the innovation to put an LED screen behind you instead of a painting? What would be the purpose and how do you use it? So how do you value the technology for the content and not the other way around: diminish the content for the technology. This is something that I think needs to be at the very center of this reflection otherwise we are just equalizing absolutely everything. Then when you are in the subway looking at an advertisement it would be the same as when you are looking at a performance because then it would be an LED screen behind. So the value of the technology is not really in the technology, it's what you are doing with the technology".*

In line with the concerns raised by the artistic director, a dramaturg emphasized the need for artistic leadership in the digitalization process. She argued: *"...I think it's important that this digital will be also driven by the artistic people. it can very easily be driven by technical people. Make them [the artistic council] decide... The most creative people who have the best ideas".*

These examples illustrate that, despite the progress of technology and the indication that the technical staff possessed particular knowledge regarding logistical and practical considerations in a digital environment, the core of artistic decision-making remained firmly with the artistic council. The artistic council, comprising individuals with substantial expertise and cultural capital, ensures that the high standards of the opera's performances are maintained. This cultural capital, as described by Bourdieu (1984), represents the accumulated cultural knowledge that confers power and legitimacy within the field of artistic production. Thus, while the technical department contributes innovative ideas regarding digitalization, their lack of equivalent cultural capital makes it less suitable for them to lead in artistic decision-making.

Aesthetics vs. Metrics. A notable aspect of negotiations, as highlighted by the interviewees, revolves around the juxtaposition of technology-enabled market orientation and an "Art for art's sake" approach to artistic orientation. With the advancements in technology, the opera has access to an abundance of consumer data. This data can be utilized to tailor the performances to meet the demands of audiences and attract more attendees. However, the artistic logic prioritizes the integrity of artistic expression, considering "art" as an abstract quality that requires no external validation (Eikhof and Haunschild, 2007). The interviews indicated that the opera navigates this duality through a continuous negotiation between the communication department and the artistic council, who initially find themselves at odds with each other. The communication department gathers and analyzes consumer data from the digital platform, Operanplay, including metrics such as the number of views, duration of views, preferences for certain performances, and general customer feedback. This data is intended to inform the opera's strategic decision and is reported to the artistic council. Although the data presents an opportunity to align offerings with audience preferences and potentially increase viewership, the artistic council may overlook its significance. As exemplified by a member of the artistic council: "*...We should not steer our whole artistic output according to what the audience wants, we should also challenge them a little bit and give them things that they didn't know* *they wanted.*" This highlights a fundamental tension, the need to balance economic considerations with maintaining cultural and artistic integrity (Lampel et al., 2000). Despite the initial disagreement, members of the communication department also recognize the importance of aligning with the artistic vision. They acknowledge that while audience metrics can guide strategic decisions, it is crucial not to lose sight of the opera's core artistic mission. A member of the communication department emphasized, "*when you're working with art it can be dangerous to fall into the trap of watching the numbers too closely, that is the hard part of the communication department. In a house where the sole purpose is to perform art sometimes what gets the most views and most selling is maybe not as important. And we can't forget that*".

These quotes exemplify the significance of disinterest in pursuing commercial gains for the sake of authenticity, as discussed by prior literature (Beverland and Ferelly 2010; Morhart et al., 2015). Despite being immersed in the market and well-versed in relevant trends, the artistic council deliberately avoids adopting a market-oriented stance. Instead, they place greater emphasis on artistic commitment and subscribe to a "l'art pour l'art" philosophy, prioritizing artistic integrity over economic considerations. Further, the congruence between the two parties regarding their shared value is consistent with Morhart et al. (2015) characterization of *integrity* as a dimension of authenticity. That is, authenticity is demonstrated when employees are intrinsically motivated by deeply held values. This is also in line with the creation of a shared worldview (Lusch and Nambisan, 2015), where cognitively distanced actors (here, the artistic council and marketing team) are brought together through a shared understanding of goals and objectives.

Digital-Phyiscal Discrepancy. The third challenge that emerged at the intersection of technology and artistic considerations involved contrasting demands of digital and physical productions. While artistic directors had become adept at working in traditional settings, the digital realm presented a new set of artistic challenges that required careful deliberation. For example, concerns were raised regarding the engagement and expression of performers, such as ballet dancers, in a digital environment. Specifically, the discrepancies between how dancers are perceived in digital recordings compared to live performances. As one artistic director emphasized: "*You can see dancers who can catch your eyes for 20 s, but in 1 h and 30 min, I mean they wouldn't have the broad spectra of artistic expression to be captivating.*" Therefore, digital recording and streaming require specialized artistic expertise to identify such discrepancies and ensure that digital recordings meet high-quality standards. In another instance, an artistic director expressed the need to hire an expert sound designer with "brilliant hearing" to detect any incorrect chords during the play. She continued by highlighting why the precision required in digital recordings is significantly higher than in live performances: "*...if you have a big orchestra of 65 people, of course, small things happen in the orchestra all the time, and you need a sound designer [who can] hear if there was a mistake...Because if you record a mistake, everybody will hear it over and over again, it's different in the physical performance where the music happens, and then it's gone*".

The above quotes showed that although the artistic directors may not participate in negotiations with another individual in this instance, the internal reflections they conduct concerning quality are deemed to be negotiations of authenticity. These reflections are crucial for maintaining the integrity and authenticity of the artistic outputs, ensuring that digital adaptations hold the same high quality as their physical counterparts. Beverland (2006) highlighted the significance of quality commitment to premium wineries' authenticity, by explaining the importance of care in selecting fruit and the intensity of care throughout the process. The meticulous actions of the artistic council, similar to the greater selectivity in raw materials by wineries, serve to further reinforce authenticity.

Social Commitment

The Royal Opera, with its substantial role in society in cultural and intellectual productions, adheres to several key social commitments, including promoting cultural education, increasing the accessibility of cultural offerings, and preserving direct engagement with its audience. Negotiations of social commitment pertained to the juxtaposition of opera's efforts to leverage digitalization as a means of meeting its cultural missions, and the consequent challenges to authenticity that stem from these efforts.

Access vs. Ephemerality. Central to its mission, the Royal Opera sought to extend its reach to a more diverse audience, ensuring that its productions were accessible to the entirety of Sweden. A producer emphasized the importance of this initiative, stating: *"As a national stage, we need to be available for all that are interested in the art forms, we need to reach younger audiences, we need to reach the whole country"*. Historically, the Royal Opera addressed this goal through nationwide tours, despite the high costs and significant logistical challenges involved. However, with the advent of online streaming, the opera found a new way to expand its outreach. By streaming performances on digital platforms, the opera made its productions available both live and on-demand to a broader audience. While digital tools facilitated an expansion in accessibility and reach, they also sparked a debate concerning their impact on the originality of the traditional art form. One producer expressed concern, noting: *"Of course, you can get a lot of really good experience in the digital world…and you can connect with a lot of institutions or artists… but in my mind, it's something completely different… it's another art form… it's not theater"*.

Furthermore, interviewees raised concerns about the challenges of capturing the emotional immediacy of live performances—feeling the performers' presence, experiencing their emotions, and interacting with the atmosphere of a live event—in a digital setting. As a choreographer pointed out: "[emotions] *are a very strong element with live performance…as a viewer, you see and feel almost intensive breathing… and you can associate it with a performer. But digital technology… can cut that, or can bring it even further"*.

The above quotes illustrate that while digital tools enabled broader dissemination of performances, the lack of emotional immediacy, phenomenological experience, or "liveness" (e.g., King, 2018; Morris, 2010), can create a sense of dissonance. This dissonance can be viewed as inauthentic, consequently implying that digital performance may hold lesser significance for opera investments.

Digital Experimentation vs. Direct Audience Engagement. The ongoing digitalization of opera not only broadened the dissemination of performances but also paved the way for the exploration of novel forms of production and artistic expression. The interviewees outlined several areas of experimenting with digitalization, such as employing digital tools to enhance in-house productions, creating hybrid performances that integrate virtual and real elements, and leveraging Virtual Reality technology to remotely control lighting, sound, and video. A particularly innovative approach outlined involved embedding performances within online gaming environments, such as the popular Fortnite gaming world. A member of the technical department described this innovative concept which represented an innovative digital experiment aimed at engaging with younger, technologically inclined audiences. He stated: *"We are moving into this Fortnite world, this gaming world... imagine an opera popping up... imagine that we have an opera house in the Fortnite world, for example. People just stop killing suddenly and enter the concert looking at performance just for a short time, and then out again and shooting"*.

However, these digital experiments require a substantial amount of time and resources, which ultimately detract from the availability of such resources for traditional performances. This presented a significant dilemma, given that opera's central mission predominantly revolved around preserving and presenting in-house performances. A producer articulated this concern: *"If we take time on stage for more rehearsals and more technical work, we lose time with the audience, and that's our mission. So, we can't just take away performances. To work more with this, we need to have the performances because we need to present something every night for a live audience"*.

Although the aforementioned quotes express opposing perspectives on the allocation of resources, they present a common interest, which is the passion for the art. Both parties present enthusiasm for the enhancement of the art form, albeit in different ways. Beverland (2009) explained this as a "love of craft" and suggested that when employees have a genuine passion for the product or service, the brand is perceived as more authentic.

Discussion and Conclusion

With this study, we explored how a cultural service addresses the dilemmas they face during digitalization, where traditions and authenticity were challenged by the adoption of new technologies. We identified three main areas of negotiations of authenticity: historical continuity (in terms of the final product and in method of production), artistic commitment (such as technical skills of performers, artistic interpretation, choreography and staging, narrative coherence, ensemble performance, and aesthetic impact), and social commitment (commitment to social goals such as promoting cultural education, increasing the accessibility of cultural offerings, and preserving direct engagement with audience). Within each of these dimensions, multiple paradoxes were negotiated, ultimately contributing to the co-creation of novel solutions and evolved views on authenticity. Notably, these negotiations took place in the social processes and practices of value co-creation; the employees at the opera did not set out to specifically improve authenticity, but authenticity was negotiated in activities such as creating the backdrop of the stage, discussions regarding customer data or deciding on the upcoming repertoire. Although each organization, from an interactionist perspective on authenticity, will have its own areas of negotiation, we expect these dimensions to be generalizable to other cultural services.

Negotiations and the Emergence of Novel Solutions

For the Royal Opera, similar to other national stages of performing arts, the pressures for digitalization had mounted for years, reaching a tipping point around 2018. From then onwards, the opera is viewed as a hybrid entity, offering both online and physical experiences. The re-institutionalization phase following this tipping point was a time of optimism and exploration, marked by aggressive digitalization visions, which were later toned down. This phase followed a dialectic pattern: the thesis (the physical opera) met the antithesis (the digital opera for everyone), leading to a synthesis (a physical opera pragmatically complemented by digital means).

In the negotiation of authenticity, each individual has their own perception of authenticity based on their personal and cultural frames (Arnould and Price, 2000) and leverages their social and cultural capital to exert influence in negotiations. Similarly, the organization's collective view of authenticity is also dynamic and influenced by various stakeholders (Arnould & Price, 2000; Beverland & Farrelly, 2010). By investigating the underlying dimensions of authenticity negotiations, we explored how new notions of values emerge, and existing ones are reinterpreted in light of digital possibilities. This perspective offers the service research a way to explore the context-sensitive nature of value creation in a collective context (Kelleher et al., 2019), as traditional markers of value, such as authenticity, are continuously redefined. Our findings show that digitalization challenged notions of the authenticity of the opera, leading to negotiations at both organizational and individual levels. Employees, deeply emotionally invested in the art form and the organization, engaged in a collective and individual learning process during this re-institutionalization phase. This period of learning and adaptation has led to temporary stability, which will inevitably be challenged as new ideas and pressures emerge both within the opera and in the surrounding service ecosystem.

The developments at the microsystem at the opera mirror the broader societal and leadership trends and reflect the values and trends of the country and the Western world at large. Digitalization is a recent phenomenon, but it is just another shift in the ongoing evolution of the opera, driven by external disturbances and internal interactions (Polese et al., 2021). In the present discourse in Western societies, an important goal is to provide access to culture for everyone, with digitalization seen as a tool to that end. A more critical interpretation would suggest that the capitalistic and neoliberal society may not be willing to provide the resources necessary for in-person cultural experiences for everyone in the same way it once was. Digitalization, therefore, serves to provide the appearance of broad access to culture without the corresponding increase in expenditure.

Additionally, while Bourdieu's conceptualization of social and cultural helps to understand the impact of power relations, it fails to account for the evolving nature of the involved parties and shifting ideas about the goal of digitalization, aspects that are made more apparent in a Hegelian dialectic analysis. In the future, as societal values and pressures shift, digital technologies will be viewed as tools to achieve other goals, continuing the cycle of adaptation and negotiation.

Digitalization and Its Influence on Value Co-Creation

By illustrating how authenticity is negotiated on the grounds of sociocultural contexts involving diverse actors and their cultural frames, we demonstrate the dialectical process of co-creation of authenticity. Similarly, we argue that the concepts of value co-creation and value can best be understood through a close focus on the deeper cultural and social meanings, as well as the interactions and dynamics involved in the process.

Digitalization represents a significant shift that not only transforms the nature of service activities (cf. Edvardsson and Tronvoll 2022) but also forces organizations and individuals to question important aspects of any service and the essence of what the service is what it is trying to achieve. Questions arise regarding who the beneficiary of the services is, what value is, and how it's co-created. More specific questions that arise may concern aspects of service, such as what customer experience the organization wants to convey, how the medium influences the core offering, and how digitalization changes the sets of competencies required. As in the case of the Royal Swedish Opera, such negotiations are likely to occur in the context of activities that ostensibly have other goals, such as byproducts.

We propose future research on how digitalization changes the perception of value and can be used as an opportunity for organizations to address fundamental issues through double-loop learning. As digitalization occurs, organizations necessarily learn how to solve things in a new medium, but with concerted reflections and discussions, a more fundamental understanding of the role of the service can be achieved (double loop learning), and how the organization can learn from change in the future (deuteron learning) (cf. Argyris and Schön, 1996).

References

Alvesson, M., Kärreman, D.: Constructing mystery: empirical matters in theory development. Acad. Manag. Rev. **32**(4), 1265–1281 (2007)

Argyris, C., Schön, D. A.: Organizational learning II: theory, method and practice. Organization development series. Boston, MA: Addison-Wesley (1996)

Arnould, E., Price, L. L.: Authenticating acts and authoritative performances: questing for self and community Collective Consumer Creativity View project Fresh Start Mindset View project (2000). https://www.researchgate.net/publication/267922936

Beerends, S., Aydin, C.: Negotiating authenticity in technological environments. Philos. Technol. **34**(4), 1665–1685 (2021). https://doi.org/10.1007/s13347-021-00480-5

Beirão, G., Patrício, L., Fisk, R.P.: Value cocreation in service ecosystems: Investigating health care at the micro, meso, and macro levels. J. Serv. Manag. **28**(2), 227–249 (2017)

Bertacchini, E., Morando, F.: The future of museums in the digital age: new models of access and use of digital collections. Int. J. Arts Manag. **15**(2), 60–72 (2011). http://hdl.handle.net/2318/105581

Beverland, M.B.: Crafting brand authenticity: the case of luxury wines. J. Manage. Stud. **42**(5), 1003–1029 (2005). https://doi.org/10.1111/j.1467-6486.2005.00530.x

Beverland, M.: The "real thing": branding authenticity in the luxury wine trade. J. Bus. Res. **59**(2), 251–258 (2006). https://doi.org/10.1016/j.jbusres.2005.04.007

Beverland, M.B., Farrelly, F.J.: The quest for authenticity in consumption: consumers' purposive choice of authentic cues to shape experienced outcomes. J. Consum. Res. **36**(5), 838–856 (2010). https://doi.org/10.1086/615047

Bourdieu, P.: The logic of practice. Stanford University Press (1990)

Bourdieu, P.: Distinction: a social critique of the judgement of taste. Harvard University Press (1984)

Braun, V., Clarke, V.: Using thematic analysis in psychology. Qual. Res. Psychol. **3**(2), 77–101 (2006)

Brodie, R.J., Löbler, H., Fehrer, J.A.: Evolution of service-dominant logic: towards a paradigm and metatheory of the market and value cocreation? Ind. Mark. Manage. **79**, 3–12 (2019). https://doi.org/10.1016/j.indmarman.2019.03.003

Campbell, D. T: Degrees of freedom" and the case study. in comparative political studies (Vol. 8, Issue 2). Sage Publications, Inc. (1975)

Cancellieri, G., Cattani, G., Ferriani, S.: Tradition as a resource: robust and radical interpretations of operatic tradition in the Italian opera industry, 1989–2011. Strateg. Manag. J. **43**(13), 2703–2741 (2022). https://doi.org/10.1002/smj.3436

Dacin, M. T., Dacin, P. A.: Traditions as institutionalized practice: implications for de-institutionalization (2007). https://www.researchgate.net/publication/242159583

De Bernardi, P., Bertello, A., Shams, S.M.R.: Logics hindering digital transformation in cultural heritage strategic management: an exploratory case study. Tour. Anal. **24**(3), 315–327 (2019). https://doi.org/10.3727/108354219X15511864843876

Dubois, A., Gadde, L.E.: Systematic combining: an abductive approach to case research. J. Bus. Res. **55**(7), 553–560 (2002)

Edvardsson, B., Tronvoll, B.,: Service management: evolution, current challenges, and opportunities, in: the palgrave handbook of service management. Springer International Publishing, pp. 35–51 (2022). https://doi.org/10.1007/978-3-030-91828-6_3

Eikhof, D.R., Haunschild, A.: For Art's Sake! artistic and economic logics in creative production. Source: J. Organ. Behav. **28**(5), 523–538 (2007).

Eisenhardt, K.M., Graebner, M.E.: Theory building from cases: opportunities and challenges. Source: J. Organ. Behav. (2007)

Evrard, Y., Krebs, A.: The authenticity of the museum experience in the digital age: the case of the Louvre. J. Cult. Econ. **42**(3), 353–363 (2018). https://doi.org/10.1007/s10824-017-9309-x

Eyerman, R., Jamison, A.: Music and social movements: mobilizing traditions in the twentieth century. Cambridge University Press (1998). https://doi.org/10.1017/CBO9780511628139

Fissi, S., Gori, E., Romolini, A., Contri, M.: Facing covid-19: the digitalization path of Opera di Santa Maria del Fiore in Florence. Eur. Plan. Stud. (2021). https://doi.org/10.1080/09654313.2021.1974352

Fraser, P.J., Fraser, I.S., Fraser, S.: Impact and OperaScotland, the listings and archive website. Arts Mark. **7**(1), 51–64 (2017). https://doi.org/10.1108/aam-08-2016-0013

Fritz, K., Schoenmueller, V., Bruhn, M.: Authenticity in branding—exploring antecedents and consequences of brand authenticity. Eur. J. Mark. **51**(2), 324–348 (2017). https://doi.org/10.1108/EJM-10-2014-0633

Giddens, A., Lash, S., Beck, U.: Reflexive modernization: politics, tradition and aesthetics in the modern social order. Stanford University Press (1994)

Gioia, D.A., Corley, K.G., Hamilton, A.L.: Seeking qualitative rigor in inductive research: notes on the Gioia methodology. Organ. Res. Methods **16**(1), 15–31 (2013). https://doi.org/10.1177/1094428112452151

Grayson, K., & Martinec, R.: Consumer perceptions of iconicity and indexicality and their influence on assessments of authentic market offerings. Artic. J. Consum. Res. (2004) https://www.researchgate.net/publication/228765554

Hanna, S., Rowley, J.: Towards a strategic place brand-management model. J. Mark. Manag. **27**(5–6), 458–476 (2011)

Hartmann, B.J., Ostberg, J.: Authenticating by re-enchantment: the discursive making of craft production. J. Mark. Manag. **29**(7–8), 882–911 (2013). https://doi.org/10.1080/0267257X.2012.732596

Hodgkiss, J., Fassio, S., Rosa, A.: Visual artists' use of digital self-marketing strategies. A quantitative survey on the present situation in Germany. Cult. Manag.: Sci. Educ. **3**(2), 53–72 (2019). https://doi.org/10.30819/cmse.3-2.03

Jones, C., Anand, N., Alvarez, J.L.: Manufactured authenticity and creative voice in cultural industries. J. Manage. Stud. **42**(5), 893–899 (2005). https://doi.org/10.1111/J.1467-6486.2005.00525.X

Keiningham, T.L., He, Z., Hillebrand, B., Jang, J., Suess, C., Wu, L.: Creating innovation that drives authenticity. J. Serv. Manag. **30**(3), 369–391 (2019). https://doi.org/10.1108/JOSM-12-2018-0383

Kelleher, C., Wilson, N., H., Macdonald, E.K., & Peppard, J.: The score is not the music: integrating experience and practice perspectives on value co-creation in collective consumption contexts. J. Serv. Res. **22**(2), 120–138 (2019)

King, T.: Streaming from stage to screen: its place in the cultural marketplace and the implication for UK arts policy. Int. J. Cult. Policy **24**(2), 220–235 (2018). https://doi.org/10.1080/10286632.2016.1150270

Lampel, J., Lant, T., Shamsie, J.: Balancing act: learning from organizing practices in cultural industries. Organ. Sci. **11**(3), 263–269 (2000). https://about.jstor.org/terms

Lusch, R.F., Nambisan, S.: Service innovation. Source MIS q. **39**(1), 155–176 (2015). https://doi.org/10.2307/26628345

Lusch, R.F., Vargo, S.L.: Service-dominant logic: Premises, perspectives, possibilities. Cambridge University Press, Cambridge (2014)

MacLean, D., MacIntosh, R.: Complex adaptive social systems: towards a theory for practice. Complex systems and evolutionary perspectives on organisations: the application of complexity theory to organisations, 149–165 (2003).

March, J.G.: Exploration and exploitation in organizational learning. Organ. Sci. **2**(1) (1991)

Maybee, J.E.: "Hegel's Dialectics", the Stanford encyclopedia of philosophy (Winter 2020 Edition), Edward N. Zalta (ed.), URL https://plato.stanford.edu/archives/win2020/entries/hegel-dialectics/.

Morhart, F., Malär, L., Guèvremont, A., Girardin, F., Grohmann, B.: Brand authenticity: an integrative framework and measurement scale. J. Consum. Psychol. **25**(2), 200–218 (2013). https://doi.org/10.1016/j.jcps.2014.11.006

Morris, C.: Digital diva: opera on video. Opera Q. 26(1), 96–119 (2010). https://doi.org/10.1093/oq/kbq002
Nowell, L.S., Norris, J.M., White, D.E., Moules, N.J.: Thematic analysis: striving to meet the trustworthiness criteria. Int. J. Qual. Methods **16**(1) (2017). https://doi.org/10.1177/1609406917733847
O'Driscoll, A.: Exploring paradox in marketing: managing ambiguity towards synthesis. J. Bus. & Ind. Mark. **23**(2), 95–104 (2008)
Orlikowski, W.J., Scott, S.V.: Liminal innovation in practice: understanding the reconfiguration of digital work in crisis. Inf. Organ. **31**(1) (2021). https://doi.org/10.1016/j.infoandorg.2021.100336
Peterson, R.A.: In search of authenticity. J. Manag. Stud. **42**(5), 1083–1098. Blackwell Publishing Ltd (2005). https://doi.org/10.1111/j.1467-6486.2005.00533.x
Polese, F., Payne, A., Frow, P., Sarno, D., Nenonen, S.: Emergence and phase transitions in service ecosystems. J. Bus. Res. **127**, 25–34 (2021). https://doi.org/10.1016/j.jbusres.2020.11.067
Russo-Spena, T., Mele, C.: "Five Co-s" in innovating: a practice-based view. J. Serv. Manag. **23**(4), 527–553 (2012)
Scott, W.R.: Institutions and organizations: ideas, interests, and identities. SAGE Publications Inc. (2014)
Smith, W.K., Lewis, M.W.: Toward a theory of paradox: a dynamic equilibrium model of organizing. In Source: Acad. Manag. Rev. **36**(2) (2011). https://about.jstor.org/terms
Södergren, J.: Brand authenticity: 25 Years of research. Int. J. Consum. Stud. **45**(4), 645–663 (2021). https://doi.org/10.1111/ijcs.12651
Trubnikova, N., Tsagareyshvili, S.: Digital challenges for creative industries: case of opera. SHS Web Conf. **114**, 01008 (2021). https://doi.org/10.1051/shsconf/202111401008
Vargo, S.L., Lusch, R.F.: Evolving to a new dominant logic for marketing. J. Mark. **68**(1), 1–17 (2004). https://doi.org/10.1509/jmkg.68.1.1.24036
Vargo, S.L., Lusch, R.F.: Institutions and axioms: an extension and update of service-dominant logic. J. Acad. Mark. Sci. **44**(1), 5–23 (2016). https://doi.org/10.1007/s11747-015-0456-3
Vargo, S.L., Wieland, H., Akaka, M.A.: Innovation through institutionalization: a service ecosystems perspective. Ind. Mark. Manage. **44**, 63–72 (2015). https://doi.org/10.1016/j.indmarman.2014.10.008
Vink, J., Koskela-Huotari, K., Tronvoll, B., Edvardsson, B., Wetter-Edman, K.: Service ecosystem design: propositions, process model, and future research agenda. J. Serv. Res. **24**(2), 168–186 (2021). https://doi.org/10.1177/1094670520952537
Wang, M.C., Lin, J.Q.-P.: The future museum shapes the museum future. Arts and the Market **8**(2), 168–181 (2018). https://doi.org/10.1108/aam-12-2017-0030
Yin, R.K.: Case study research and applications (2018)
Zeithaml, V.A., Verleye, K., Hatak, I., Koller, M., Zauner, A.: Three decades of customer value research: paradigmatic roots and future research avenues. J. Serv. Res. **23**(4), 409–432 (2020). https://doi.org/10.1177/1094670520948134

Saba Samadilashkariani is a Ph.D. candidate and course instructor at Stockholm Business School, Stockholm University. Her research focuses on digital transformation in cultural industries and the influence of digitalization on physical service aspects.

Jon Engström is Assistant Professor in Marketing and Head of the Marketing Section at Stockholm Business School, Stockholm University. He obtained his Ph.D. at Linköping University (Sweden). His research concerned user behavior and experience, in the context of service development and often in relation to health and healthcare. His recent research emphasizes the role of technology and how it influences users, business and the arts.

Chapter 11
Towards a 'Re_cognition Income': Addressing Wealth Reallocation in the Digital Age

Sergio Barile and Francesca Iandolo

Abstract This manuscript addresses the socio-economic challenges arising from technological advancements, focusing on their role in widening wealth inequality. It introduces the concept of 'Re_cognition Income' as a mechanism for fair wealth redistribution, aimed at fostering social equity in a rapidly evolving digital landscape. By examining historical shifts from Proprietary to Financial Capitalism and the emerging Participatory Capitalism paradigm, the study provides a comprehensive framework for understanding these transformations. The methodology integrates historical analysis with socio-economic theory to identify key triggers of paradigm shifts and their implications for wealth distribution. Through this approach, the study aims to support policymakers in formulating strategies that ensure the inclusive sharing of technological benefits, promoting a sustainable and just society.

Keywords Technological paradigms · Socio-economic transformation · Re_cognition income

Introduction

The contemporary international media landscape is increasingly dominated by continual information on technological innovations transforming organizational processes across all sectors, both public and private (Castells 2010; Brynjolfsson & McAfee 2014). These advancements are revolutionizing individual and collective activities within various dynamics. Ongoing discoveries and technological developments are significantly reshaping our daily lives, work environments, and interpersonal and inter-organizational relationships (Castells 2010; Rifkin 2014; Tapscott

S. Barile · F. Iandolo (✉)
Department of Management, Sapienza University of Rome, Rome, Italy
e-mail: francesca.iandolo@uniroma1.it

S. Barile
e-mail: sergio.barile@uniroma1.it

D. Sarno et al. (eds.), *Complexity and Emergence in Market Ecosystems*, New Economic Windows, https://doi.org/10.1007/978-3-031-81942-1_11

2015). This relentless progression elicits both enthusiasm for the opportunities it presents and concerns regarding its ethical, social, and economic implications.

Concurrently, there is an observable escalation in wealth inequality (Piketty 2014). While certain individuals and organizations benefit substantially from technological advancements, improving their efficiency and performance, a considerable segment of the population struggles to keep pace, resulting in marginalization or impoverishment (Stiglitz 2012; Klein 2008). This asymmetric distribution of wealth is already engendering social tensions, prompting debates on critical issues such as social justice, equity, and sustainability. The intricate dynamics generated by accelerated technological innovation and wealth disparity present a scenario that is challenging to interpret and predict (Foucault 1995; Habermas 1984; Deleuze & Guattari 1987; Vargo et al. 2023; Polese et al. 2018).

Addressing these challenges equitably and effectively, ensuring that the benefits of technological development are widely shared, is imperative. How can we foster a more inclusive and sustainable society that elevates marginalized populations to an adequate standard of living? Is it possible to create a comprehensive representation that integrates various informational aspects of current events, thereby providing a clearer understanding of the scenario?

This chapter proposes a conceptual framework that integrates dimensions related to Technological Innovations, Wealth Distribution, Social Impacts, and Sustainability. This framework aims not only to elucidate the observed paradigm shifts but also to support policymakers in addressing contemporary challenges effectively by offering an interconnected perspective of current socio-economic phenomena.

A crucial aspect for understanding the evolutionary mode characterizing the succession of paradigmatic cycles is the identification of events that can be classified as triggers for change. An analysis, even if not in-depth, of the historical periods associated with circumstances indicating a paradigmatic shift reveals undeniable correlations with scientific discoveries and innovations. A closer examination of which discoveries and innovations have had the most significant impact in determining paradigm changes in each period incontrovertibly shows that those capable of permanently altering human conceptions of time and space hold absolute significance.

Paradigmatically significant innovations often radically change human understanding and perception of the surrounding world (Polanyi 2001; Schumpeter 2008; Marx 1992). These shifts in the conception of time and space have profound cultural and social implications. They influence all existential contexts: how people work, communicate, study, and entertain themselves. Consider how the digital revolution and the proliferation of mobile devices have made it possible to communicate, work, and access information from almost anywhere at any time, thereby completely redefining our perception of space and time. In summary, it is reasonable to assert that true paradigm-shifting innovations are those that change how we perceive and interact with the world.

In this study, we identify three main paradigms—Proprietary Capitalism, Possession (or Financial) Capitalism, and Participatory (or Knowledge) Capitalism—each representing distinct socio-economic structures influenced by technological and philosophical developments. These paradigms were identified through a historical analysis of key events and innovations that reshaped human perceptions of time, space, and value. In Proprietary Capitalism, value was rooted in tangible ownership, such as land and physical assets. In the Financial Capitalism paradigm, value shifted toward the accumulation of capital and financial instruments, where wealth could grow through representations rather than physical goods. In Participatory Capitalism, value is increasingly associated with digital engagement and community participation within virtual ecosystems. While the paradigms may overlap and coexist due to varying individual and collective perceptions, each one is associated with specific timeframes: Proprietary Capitalism generally spans from the early industrial age through the late nineteenth century, Financial Capitalism emerges in the twentieth century, and Participatory Capitalism is a paradigm of the digital and information age, gaining prominence in the twenty-first century.

To illustrate, consider the example of a teenager who spends significant time playing video games in an online ecosystem, contributing to the game's community by creating content, interacting with other players, or achieving high scores that enhance the game's appeal. Traditionally, this kind of engagement might be seen as "leisure" or unproductive time. However, in Participatory Capitalism, the teenager's interactions add value to the platform by generating data, sustaining player communities, and even drawing in new users. Recognition Income could provide a structured means of compensating these contributions, acknowledging the value that participants create even in digital environments that traditionally go unrewarded. Such a model not only recognizes the importance of digital engagement but also encourages users to develop skills and expand their involvement in state-recognized digital activities.

In the following sections, based on the developed representation, we proceed with a historical description of the socio-economic aspects of the period referred to as the Proprietary Capitalism Paradigm, followed by the Possession Capitalism Paradigm, and then culminating in a description of what the authors consider the imminent Participatory Capitalism Paradigm. This historical perspective aims to provide evidence of what has occurred and what is currently unfolding within these paradigms.

Historical Contexts and Paradigm Shifts

The Paradigm of Proprietary Capitalism

Technological change, new philosophical sensibilities, and specific historical events have contributed to the rise of ever-expanding manufacturing activities, characterizing a period known as proto industrialization (Hobsbawm 1999; Mokyr 1990). During this period, the production of goods began to move away from rural homes and develop in separate workplaces, gradually and inexorably replacing previous physiocratic-feudal systems. This shift marked the emergence of new social figures and elites. The rise of industrialization and the transition from agrarian to manufacturing economies laid the groundwork for what we identify as the Paradigm of Proprietary Capitalism. This paradigm is characterized by a distinct relationship between labor, ownership, and social structures that emerged alongside technological and political transformations.

Figure 11.1 highlights the three primary drivers of this paradigm: government, management, and elicitation.

Government established the rules, protecting private property and incentivizing economic growth through supportive laws. Management was responsible for organizing labor and resources, focusing on productivity within factory systems. Elicitation, in this context, refers to the state and social systems mobilizing citizens toward productive roles in the economy, where ownership and wealth were seen as paths to stability and social status.

Proprietary capitalism is thus characterized as a socio-economic domain where individuals and enterprises are incentivized to own and control resources, production, and capital.

Historically, a growing divergence was observed between the large labor forces and the centralized decision-making processes controlled by technical and economic elites within companies (Tilly 1990; Landes 1969). This disparity in knowledge, skills, and abilities across levels of industrial organizations (Braverman

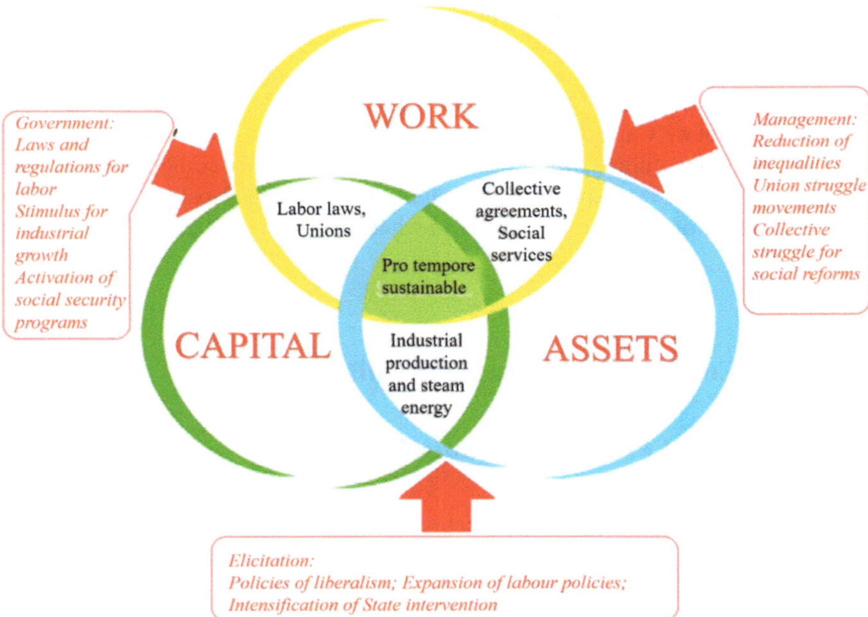

Fig. 11.1 The paradigm of proprietary capitalism. *Source* Authors' elaboration

1974; Thompson 1963) highlighted the distinct hierarchies that defined Proprietary Capitalism.

From a social perspective, the second industrial revolution marked an expansion in rights, characterized by a significant reduction in inequality and the gap between social classes due to rising incomes (Chandler 1977; Hobsbawm 1987). Property ownership became a primary measure of individual success, providing financial security, social recognition, and stability, although sometimes disrupted by cyclical epidemics (Weber 1930; Tocqueville 2003). The concept of property as value tied economic success to ownership, creating social stratifications based on material wealth.

Alongside these socio-economic developments, politically the rise of Proprietary Capitalism coincided with the emergence of the nation-state emerged (Anderson 1083; Tilly 1975). Originating from absolutism and influenced by romantic ideals and various identity aspirations, the nation-state eventually replaced old dynastic regimes by the nineteenth century.

The primacy of politics was integral to this paradigm, as sovereign states became the ultimate rule-making authorities. Sovereignty is defined as a 'supreme, legally independent, and non-derived power' (Bodin 1992; Schmitt 2007). Only politics, rather than administration, could act as the 'final decision-maker' and set priorities.

Even when, after World War II, the 'muscular bipolarity' between the USA and USSR forced nation-states to partially cede sovereignty in foreign policy to their macro-blocs (bringing the world back to a quasi-pre-Westphalian era), the primacy of politics in domestic policy was preserved (Judt 2005; Gaddis 2005). This preservation was ensured by the Schmittian necessity for each state to choose between 'friend' and 'foe,' allowing for the retention of substantial 'sovereign power' despite international constraints.

In Proprietary Capitalism, the economy revolved around labor and wages. Many scholars argue that the issues related to the tendency for capital to concentrate in the hands of operators driven by an imperative for dimensional growth can be resolved by redistributing part of the surplus value (in the Marxist sense) appropriated by capitalists, leveraging continuous technological progress (Marx 1992; Russo 2019). This approach was intended to mitigate economic disparities and sustain social stability.

Proprietary Capitalism thus emerges as a socio-economic domain where ownership of resources, production, and capital is paramount. Wealth and status are tied to material assets, and the paradigm is structured around a rigid perception of time and space: time is regulated through industrial schedules, while space is defined by ownership of tangible assets like land and factories.

The Paradigm of Possession—or Financial Capitalism

As industrialization gave rise to Proprietary Capitalism, shifts in global trade, technology, and finance led to a new phase in the late twentieth century, identified as the Paradigm of Possession or Financial Capitalism.

From the early 1990s, historical events facilitated the formation of a genuine 'financial industry,' which expanded by leveraging information markets to virtually multiply its scale. Financial instruments, especially derivatives, became the primary means of wealth generation and economic progress, diverging significantly from the traditional real economy (Helleeiner 1994; Lipuma & Lee 2004; Bryan & Rafferty 2006; MacKenzie 2006).

The fall of the Berlin Wall in 1989 led to a progressive opening of international trade, accompanied by a steady reduction in the role of states and a shift away from decentralized manufacturing (Stiglitz 2002; Gowan 1999; Gaddis 2005; Judt 2005). Production increasingly moved to areas with less concern for labor quality, pollution, and health, giving rise to a new form of finance disconnected from the real economy (Arrighi 1994; Krippner 2011). This era marked the beginning of a 'liquid society' (Bauman 2000; Lury 2011; Sennett 2006), where affiliation was no longer grounded in ideology but in individual participation in consumption. This is the onset of what is termed the Paradigm of Possession or Financial Capitalism (Fig. 11.2).

Figure 11.2 highlights the three primary drivers of Financial Capitalism: consumptions, finance, and debt capacity. This economic model, evolving from Proprietary Capitalism amidst growing globalization and the decline of manufacturing industries, rests on the premise that wealth generation and economic progress primarily occur through derivative financial activities. Thus, institutions such as banks, investment

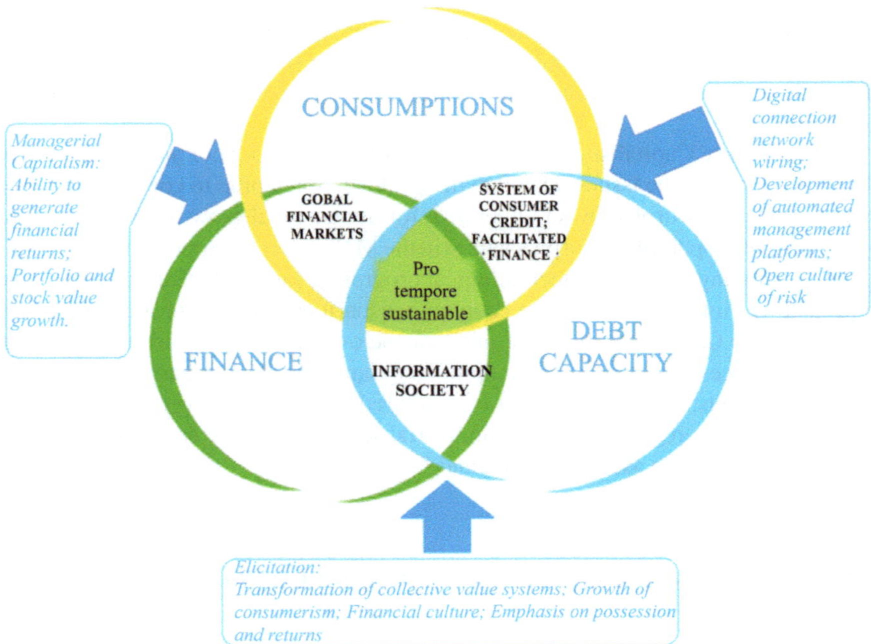

Fig. 11.2 The paradigm of possession. *Source* Authors' elaboration

firms, and hedge funds play a crucial role as facilitators and mediators in capital accumulation processes. The 'value' previously associated with ownership transforms into the availability and use of assets and services. The measures of individual quality and success shift to metrics of creditworthiness and debt capacity as recognized by the financial system (Stiglitz 2010; Harvey 2005; Rajan 2000; Krippner 2011). Belonging to the financial system fulfills security and status needs, with social recognition linked to credit access and debt capabilities (Langley 2008; Poon 2009; Piketty 2014; Graeber 2011).

The United States and the United Kingdom were among the first to undergo this transformation, establishing themselves as key players in a globalized economy dominated by financial capital flows. Money transcends its role as a mere medium of exchange to become an entity that can be traded, exploited, and manipulated for profit. High-profile corporate defaults in the early twenty-first century, such as Enron and Lehman Brothers, triggered a shift, positioning financial operators as key players in socio-economic transformation. A concerning feature of renewed financial capitalism is the influence of its components on Western political dynamics. This influence is exerted through political lobbying and the implicit threat of speculative attacks on national economies, promoting an ethos centered on profit maximization, eroding values of solidarity, cooperation, and merit-based respect (Ferguson 1995; Hacker and Pierson 2010; Davis 2009). During the Third Industrial Revolution, economic globalization and technological innovation jointly redefined the role of the nation-state. This phenomenon, marked by the rise of multinational corporations and intensified transnational capital and service flows, was energized by information technologies, creating a complex and interconnected international landscape (Castells 2010; Friedman 2005). Traditional national boundaries began to blur, projecting economic and social issues onto a global stage populated by significant non-institutional actors. Financial market liberalization and multinational corporations further extended the reach of finance, pushing loan volumes and enabling widespread deferred payment systems, intensifying global economic dependence on finance (Lazonick & O'Sullivan 2000; Tooze 2018).

This expansion led to increased indebtedness among governments, corporations, and citizens, often limiting the autonomy of nation-states heavily reliant on international finance. The political landscape faced the rise of new powerful actors: technological governance entities, often with greater wealth and influence than many states, challenging the authority and regulatory capacity of nation-states (Zuboff 2019; Srnicek 2017; Van Dijck et al. 2018). In this reconfiguration, nation-states navigated uncharted waters, struggling to balance the needs of their citizens with the demands of an increasingly globalized economy. The peak of globalization posed deeper challenges to state sovereignty, progressively expropriated by technology giants.

Different governance dynamics emerged, no longer responsive to state authority but conditioned by supranational 'governances,' manifesting through the consolidation of rules, principles, and procedures aimed at managing and directing collective phenomena (Rodrik 2011; Habermas 1998). National systems, driven by 'corporate' logics, increasingly heeded the priorities of elite groups and lobbies, rather than popular will, influenced by finance and markets, which naturally disregard national borders and collective interests.

The Paradigm of Participation

Following the paradigm shift to Financial Capitalism, with its emphasis on asset possession and creditworthiness, technological advancements and the expansion of digital platforms have led to a new transformative phase: Participatory Capitalism. This paradigm represents a shift toward value creation through continuous digital interaction and engagement, establishing a more immersive, user-driven economy.

Each transformation in the value chain, irrespective of the factors initiating the change, does not merely replicate previous dynamics at different levels. Instead, it defines entirely new behavioral patterns and systemic logics, thereby reshaping economic and social processes and delineating the boundaries of a new paradigmatic framework (Rodrik 2011; Habermas 1998). This results in a set of trends that lead to a rapid transition towards a system grounded in Financial Capitalism but increasingly digitalized, featuring higher levels of efficiency, accessibility, and innovation, albeit simplified in options and increasingly standardized (Benkler 2006; Tapscott 2015).

Figure 11.3 highlights the primary drivers of the Paradigm of Participation: consumption, digital interactions, and platform engagement. In this paradigm, economic and social value is generated by continuous user participation and data-driven interactions.

Financial hegemonies, in their effort to enhance their control and governance over major economic flows, inadvertently foster the development of digital processes organized on information platforms (Brynjolfsson and McAfee 2014). This causes the role and freedoms of human action, and thus their organizational systems, to be progressively deconstructed and compressed into platform logics, becoming standardized. The unintended consequence of this trend leads to a complete revolution of organizational processes across various productive domains (Ford 2015). Within a short time, scenarios arise where any issue, not just financial, is addressed with solutions that inevitably lead to further growth in the ‘density’ of computerized activities within operational processes. This applies to all organizational processes: public and private, profit and non-profit, service and production companies, various professionals, and even private activities of individual citizens.

A less evident consequence of this ‘dehumanization’ of activities within processes is the transfer of knowledge (Polanyi 1966; Nonaka and Takeuchi 1995; Davenport and Prusak 2000). Knowledge, understood as the cumulative result of practices, schemas, models, and information derived from problem-solving experiences and observations, accumulated over centuries of social and civil consolidation and cultural development, is transferred into increasingly advanced and efficient algorithms and platforms (Drucker 1993; Bell 1973; Castells 2010; Porter 1985; Teece 2009; Chesbrough 2003).

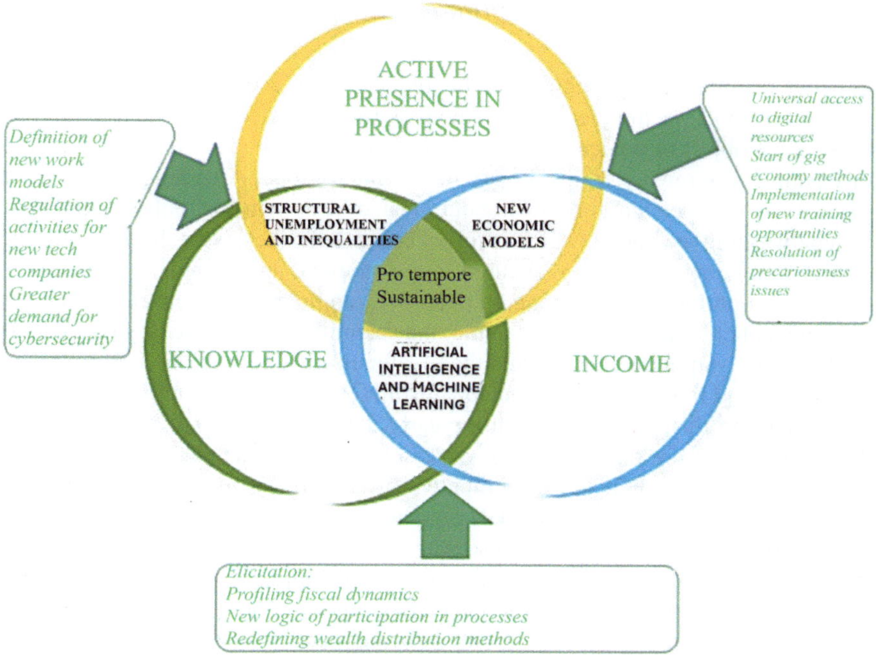

Fig. 11.3 The paradigm of participation. *Source* Authors' elaboration

This assertion requires graphic emphasis because, despite its seemingly obvious and universally acceptable nature, it introduces a significant factor. It encompasses elements that help clearly understand otherwise obscure aspects, which, when integrated, provide a new perspective and, it is hoped, greater awareness.

In the overlapping areas of Fig. 11.3, terms like "pro-tempore sustainable" reflect the concept of temporary sustainability, achieved through constant digital engagement and consumption. This notion of sustainability relies on continuous interaction, where value is sustained by user presence in digital ecosystems. Moreover, in the digital era of the Paradigm of Participation, perceptions of time and space are fundamentally redefined. AI-driven automation and virtual environments compress time, enabling real-time interactions and decision-making, making tasks feel immediate and seamless. Space, meanwhile, becomes increasingly virtual, with many interactions occurring in digital "metaverses." This allows individuals to transcend physical boundaries, creating new perceptions of proximity and presence in virtual spaces.

The Notions of Method, Technique, Tool

To better understand the significance of this assertion, a brief digression into the concepts of tools, techniques, and methods is useful (Barile et al. 2017a, 2017b). A tool is an object or device designed to perform a specific function within a task, for example, a screwdriver, a software tool, a microscope. A technique refers to a specific way or procedure for carrying out a task or activity using tools; for example, the handwashing technique, which involves using tools like a water basin, soap, and a towel in a known manner. A method is a systematic and organized approach to addressing a problem or achieving an objective, employing various techniques and tools in a coherent process; for example, the scientific method used by researchers involves observation (techniques and tools), hypothesis formulation, data collection (techniques and tools), analysis, and conclusion formulation (techniques and tools).

Considering this, we note that in the early days of computing, the general use of systems (hardware plus software) was essentially instrumental. To better understand, consider the initial use of computers by notaries. Undoubtedly, in the first years following the launch of personal computers, notaries used the tool as a replacement for traditional typewriters. Those who remember that period know that many preferred using 'enhanced' typewriters over actual computers. Only later, when the expanded use of the tool led to the understanding that many documents (notarial deeds) could be stored and modified without being redrafted each time, did the procedure change, and the computer gained 'technical' value. More recently, the ability to have 'online' legal collections interacting in real-time with document drafting, along with predefined templates for specific types of deeds, has elevated the Information System (IS) to a 'method.'

From this example, several considerations arise. In the instrumental phase, the IS could not acquire any knowledge, such as the 'ability to draft a deed,' which remained the exclusive prerogative of the notary (a human). In the technical phase, through the involvement of IT technicians alongside innovative-oriented notaries, part of the legal knowledge, generally limited to what the notarial office decided to computerize, was incorporated into the IS. In the methodological phase, much legal knowledge, sometimes exceeding that of less equipped notaries, is available in the IS. It is not far-fetched to think that an individual without a law degree but proficient in notarial procedure software could produce deeds as effective as those of qualified notaries lacking the digital enhancement.

This concept extends to any professional category, including engineers, lawyers, accountants, and even judges, who use IT support in their activities. In the technical phase, the knowledge input was limited to contributions from professionals adopting an IS, considered their intellectual property (with restrictions on transferring developed models to potential competitors). In the methodological phase, who should be considered the legitimate owner of the knowledge included in the IS? In fact, the knowledge possessed by the IS has become an entire compendium of legal knowledge, continuously updated and now even enhanced by artificial intelligence!

Labor, Wealth Redistribution, and Value in Participatory Capitalism

A market survey would likely reveal that many people believe the collective sum of knowledge incorporated in any IS belongs to everyone, making it public property. However, this hasty evaluation introduces a significant perspective error: the knowledge indeed belongs to those in the physical world ('World 1'), not to entities and their representatives in the digital world ('World 2'). "World 1" refers to the material, physical world, where traditional economic activities are based on tangible assets and direct interactions. In contrast, "World 2" represents the digital domain, consisting of data, algorithms, and processes driven by code rather than physical resources. In this virtual world, economic value emerges from digital engagement, where algorithms, platforms, and artificial intelligence redefine how wealth and influence are created (Fig. 11.4).

Over time, this appropriation of resources from established knowledge, like manuals or intellectual or artistic works, leads to their removal from the real economy and transfer to a 'non-place,' monopolized by a narrow and exclusive group increasingly holding global wealth, to the detriment of individuals (holders of 'potential value,' a product of centuries of civilization, always accessible through simple 'participation') (Arthur 2017; Benkler 2006; Zuboff 2019).

The transfer of value from the real economy to the digital economy is a phenomenon characterized by various interaction schemes, made more complex by its ongoing evolution. Accurately accounting for the actual amount of resources taken from 'World 1' and transferred to 'World 2' requires reformulating the value chain hypotheses for products and services as shared and consolidated in 'World 2.'

The call for politics to ensure that the community enjoys a prosperous yet orderly and 'natural' economy, shielded from the potential perversions of crematistics,

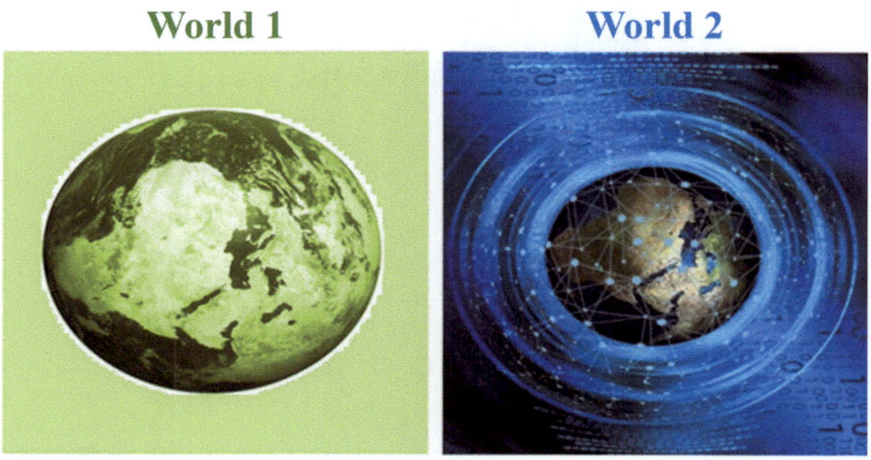

Fig. 11.4 'World 1' and 'World 2'. *Source* Authors' elaboration

protecting every community member from the obsessions of personal enrichment, is more relevant than ever.

The immense wealth generated by 'crematistic' operators (digital world giants) in the 'third paradigmatic model,' termed 'Participatory Capitalism,' is largely based on free utilizations (or, in some cases, outright appropriations) of established cultural, material, and immaterial excellences from 'World 1.' The concentration of wealth increasingly in the hands of a few has reached unjustifiable levels (Fig. 11.5).

Bill Gates, nicknamed 'Farmer Bill' by Land Report, is now America's largest private farmland owner, with 300,000 acres in 19 states. Mark Zuckerberg has announced his new project of raising livestock in Kauai, Hawaii, aiming to produce high-quality meat. Jeff Bezos is constructing a colossal 42-million-dollar clock in a Texas mountain, designed to chime every millennium, intended to function for 10,000 years even in the absence of humans.

As illustrated, many infrastructural, cultural, architectural, artistic, and natural elements from 'World 1' are exploited by 'World 2.' For example, consider the arts, professions, customs, and cultural advancements, including artworks, intellectual works, architectural and landscape structures, and socio-cultural phenomena produced by the 'real world' ('World 1'). A moment's reflection reveals that infrastructures like cabling necessary for the internet, computer literacy provided by public education, and countless scientific and cultural information, images of streets, artworks, monuments reproduced millions of times daily, and historical traditions

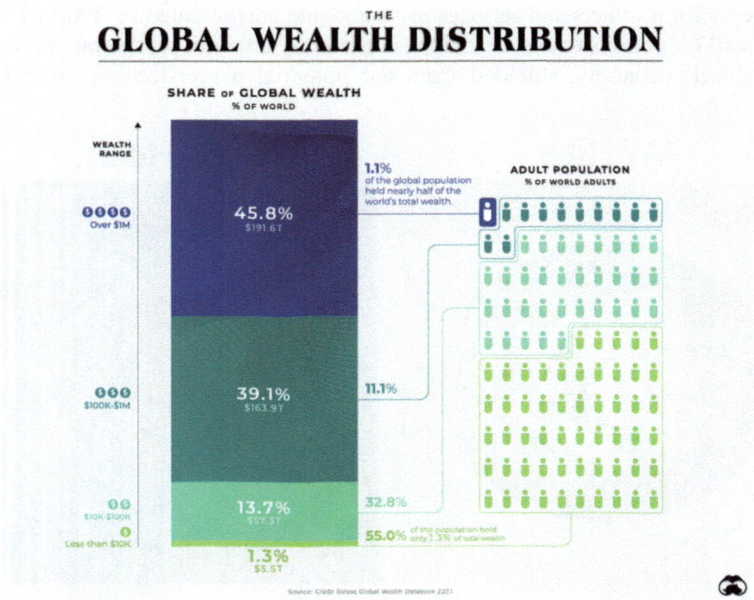

Fig. 11.5 Global wealth distribution. *Source* Credit suisse 2021

are essential elements for the digital world, offered freely through voluntary and involuntary participation processes.

Therefore, it is indispensable for the state, exercising real sovereignty, to impose criteria and rules on the significant and minor operators of 'World 2' to 'compensate' 'World 1.' The state could impose a fee proportional to the value generated by its citizens' participation in 'World 2,' demand a price for the 'raw material' (knowledge from 'World 1') or require royalties on products and services using 'World 1' knowledge, images, and scientific results. This approach could not only fund public expenses but also address a significant societal issue exacerbated by the dominance of 'World 2' over 'World 1': the inability to redistribute wealth through the now obsolete 'labor' driver.

The advancement of cutting-edge technologies in robotics and artificial intelligence drives economic dynamics towards a condition where non-human entities increasingly undertake various activities and processes, impacting sectors from retail and media to financial services and digital health (Fig. 11.6). Many professional roles risk disappearing: 40% in emerging economies, 26% in low-income countries, and up to 60% in advanced economies.

It is reasonable to anticipate that within a few years, large segments of the population will be excluded from employment, necessitating new forms of income redistribution. Several Western countries already have programs supporting financially vulnerable individuals or families, such as Italy's citizenship income, Finland's Universal Basic Income (UBI), Germany's Hartz IV, Spain's Seguro Social de Ingreso Mínimo (SSIM), and the US Welfare System. However, these measures, lacking a

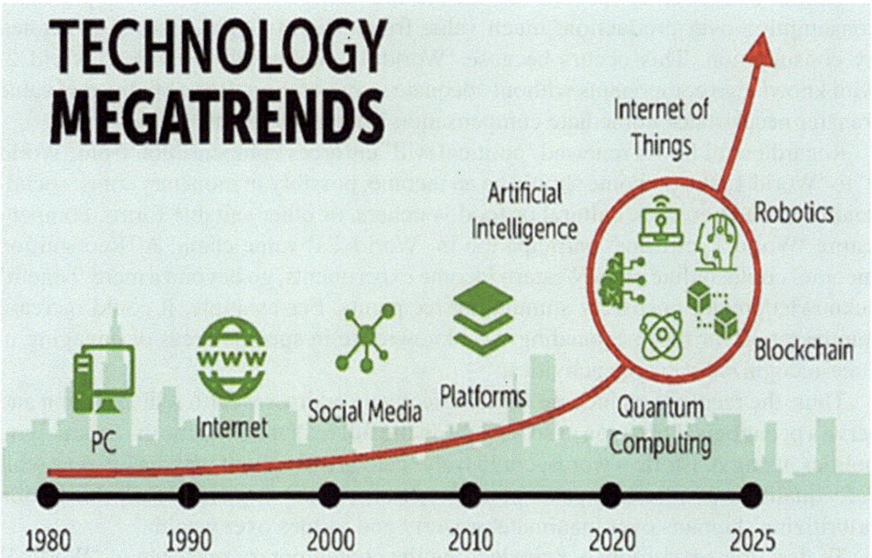

Fig. 11.6 Emerging technology trends. *Source* Hackermoon.com

systemic vision of the current paradigm, have generally produced fewer effective results, both materially and morally.

Without a strategic plan for wealth reallocation, these measures appear as tactical remedies, generating simplistic perceptions that working, tax-paying citizens fund others to live in idleness. Such interpretive drift is plausible, especially if income recipients are expected to reintegrate into the labor market. However, the necessary measures should guarantee state support for those excluded from the labor market due to the declining need for human labor. Previous attempts addressed issues emerging with a new paradigm but lacked an evolutionary systemic vision, resulting in ineffective solutions based on outdated paradigms.

A state capable of exercising true sovereignty could impose 'reparations' from 'World 2' to 'World 1' for the value appropriated daily. This could establish a 'recognition income,' remunerating the 'knowledge producer' role of every 'World 1' citizen integrated into 'World 2's' value chain. In the first paradigm, value was material, linked to ownership (land, buildings, raw materials, products). In the second paradigm, value was based on representations (stocks, bonds, derivatives, multiproperty, renting). In the third paradigm, value evolves into a new 'significance' within 'World 2,' where most virtual activities, services, and platform-based operations now occur. This shift results in a dematerialization process, with many interactions once involving 'World 1' components now replaced by 'World 2' components. Most daily interactions are already human/platform, with algorithmic logic prevailing over human logic.

The ethical impact of this trend is profound, but outside the scope of this discussion. What matters here is that individuals, in various roles, contribute to service utilization and asset use processes. For the first time, economic metrics prioritize consumption over production: much value from 'World 2' processes is generated by consumption. This occurs because 'World 1' unwittingly provides 'World 2' with knowledge components without adequate remuneration. Recognizing this value transfer necessitates immediate compensation acknowledgment.

Regardless of how a renewed 'political will' enforces compensation from 'World 2' to 'World 1,' the outcome should be an income, possibly in monetary sums, social-health service booklets, cultural or food vouchers, or other suitable forms, compensating 'World 1' citizens' participation in 'World 2's' value chain. A 'Recognition Income' could, unlike other Western income experiments, go beyond a mere 'benefit' acknowledgment, positively stimulating recipients. For example, it could increase remuneration for those expanding their knowledge in specific areas or engaging in state-recognized valuable activities.

Thus, the recognition income would effectively address wealth redistribution and serve a pedagogical function in society. With the full realization of the third paradigm, politics, using Aristotle's words, could truly 'care greatly about making citizens ethically qualified, good, and capable of noble actions,' achieving orderly administration prioritizing 'humans over inanimate property and virtues over wealth.'

To describe a paradigm acknowledging the contemporary presence of 'World 1' and 'World 2,' we must recall the foundational aspects of the socio-economic context. 'Industry 4.0,' underpinning 'participatory capitalism,' is defined by strong product and service customization and self-organizing ecosystems, ensuring a close link between the real world (workers, machines, raw materials) and the virtual world. The novelty lies in using cyber-physical systems connecting objects and people through global information and relational networks, employing key innovative technologies (IoT, Big Data, 3D printing, AI, bioengineering, cloud computing, platforms).

Artificial intelligence (AI) and machine learning significantly impact time and space perception, allowing faster information processing and instant access to information and services, reducing the perceived time for tasks. AI can automate repetitive tasks, reducing time spent on routine activities. AI and virtual reality create increasingly immersive virtual experiences, allowing users to engage in digital worlds with avatars, significantly altering space and time perception. People can participate in virtual events, travel to virtual locations, or perform activities in virtual environments, often immersed in multiple 'metaverses' simultaneously.

Hence, AI and machine learning have the potential to generate inequalities and increase unemployment. We stand at a dramatic crossroads: the digital age seems to signal the end of established models (mental, educational, organizational, cultural, social) while demanding new sensory capabilities for socio-economic reconfiguration. This social innovation is slow to manifest.

Technological development, as long as it qualifies instrumental and technical innovations, confines the value produced within the material world ('World 1'). Scientific innovations in their instrumental or technical phase produce technological impacts used in specific activities within processes, remaining subservient to human value creation strategies.

As Brian Arthur (Arthur 2017) and other scholars noted, even when technology remains instrumental and technical, circumscribed within relevant 'World 1' processes, the complete amortization of code costs still results in value dispersion into 'World 2'. 'World 2' comprises data, information, processes, and interactions through electronic devices and digital networks, particularly codes, algorithms, and platforms increasingly impacting value creation processes.

In conclusion, understanding these dynamics emphasizes the urgent need for state-imposed compensation criteria and rules from 'World 2' to 'World 1,' ensuring fair value distribution and addressing wealth redistribution challenges in the face of decreasing labor necessity due to technological advancements.

The Re_cognition Income

What is fundamentally occurring is that the value previously generated by processes entirely internal to 'World 1' is progressively and at an increasing rate, transferring to 'World 2'. As Brian Arthur prophesied, "This is bringing the economy into a new and different era. The economy has reached a point where it produces enough for everyone, but the means of accessing these services and products, and jobs, are constantly narrowing. Therefore, the new period we are entering is not so much about production, about what is produced; it is about distribution: how people get a share of what is produced. Everything, from trade policies to government projects to business regulations, will be judged by distribution in the future. Politics will change, free-market beliefs will change, social structures will change."

Brian Arthur's reflections on technology and economic evolution provide a foundational perspective on the shift in value creation from traditional production to digital and service-oriented processes. In line with Service-Dominant Logic (SDL), which views value as co-created through interactions and service exchanges rather than embedded in tangible products, Arthur's ideas resonate strongly within the Participatory Capitalism paradigm (Vargo & Lusch 2004, 2008; Lusch et al. 2017; Vargo et al. 2023). SDL emphasizes that value is realized not solely through ownership but through active participation and the fulfillment of specific needs within networks, an idea that aligns with Arthur's observation of technology as an enabler of decentralized, adaptive systems. In the context of World 2, this approach redefines wealth not by material accumulation but by the facilitation of meaningful, service-based interactions across digital platforms. By integrating Arthur's perspective with SDL, we gain a clearer understanding of how digital engagement fosters a new type of value creation—one that is flexible, networked, and increasingly shaped by algorithms and data-driven interactions (Barile et al. 2013; Golinelli et al. 2010). This paradigm shift underscores the need for Recognition Income as a means to fairly compensate individuals whose participation in these digital networks contributes to the co-creation of value within the economy of World 2.

The dynamic with which superhuman entities (algorithms and platforms) are replacing activities previously the prerogative of humans is growing exponentially. Nearly half, precisely 49%, of the jobs currently performed by humans globally could be automated when "currently developed" technologies are deployed on a global scale. This is revealed by a study by McKinsey Global Institute, which takes into account not just individual jobs (e.g., 'farmer,' 'manufacturing worker,' 'IT technician') but specific tasks performed ('agricultural machinery operator,' 'turning machine operator,' 'system administrator'). It examines 54 countries, accounting for about 78% of the world's workforce (Sarcina 2017).

The dynamics of various attempts and methods of income allocation developed by Western countries can vary even within the same country depending on regional and local policies. Each method adopted faces difficulties related to financial sustainability, eligibility criteria definition, complexity of controls, and the disincentive impact on job-seeking availability. Many nations continue to experiment and adapt such programs to address the specific challenges of their socio-economic contexts.

The future scenario, based on the described transfer of most value-creation activities from 'World 1' to 'World 2', will set the stage for increasingly unbalanced wealth allocation (Fig. 11.7).

It is only a matter of time before artificial intelligence revolutionizes the global economy, but it risks overwhelming many jobs worldwide. Moreover, AI is almost certain to increase social inequalities and widen the gap between richer and poorer countries.

Fig. 11.7 The draining of resources. *Source* Authors' elaboration

Societies in 'World 2' will witness an exponential increase in their wealth, and attempts to mitigate this through philanthropy, sponsorships, charity, and other voluntary support forms from 'World 2' to 'World 1' will be insufficient. The evolution of support methods for vulnerable groups will depend on political priorities and the ability to enable resource restitution through emerging challenges in each country. The 'elicitation' phase must prioritize defining a revolutionary method of human participation in wealth created by superhuman entities (algorithms and platforms), currently marked by enormous distributive inequalities (Fig. 11.8).

Criteria and methods must be defined to determine how much of the value created in 'World 2' should benefit 'World 1,' how this wealth should be recognized for the 'World 1' community, and how much should be attributed to individuals. The issue of reducing inequality and improving the well-being of the most vulnerable will likely remain a central concern in political and social debates (Benanti 2018; Benanti 2020; Chiriatti 2019).

Notably, the primary characteristic of 'World 2'—the objective traceability of each actor's contribution (whether human, superhuman, or posthuman) to individual activities—becomes a useful principle for attributing a portion of the wealth produced.

Thus, a hypothetical allocation based on weighted participation in processes becomes the appropriate method for wealth distribution. For example, even a teenager playing video games, through their participation as a user, would earn the right to a share of the value created by the overall process. We will no longer have idlers or unemployed individuals, as long as each individual can participate in a specific

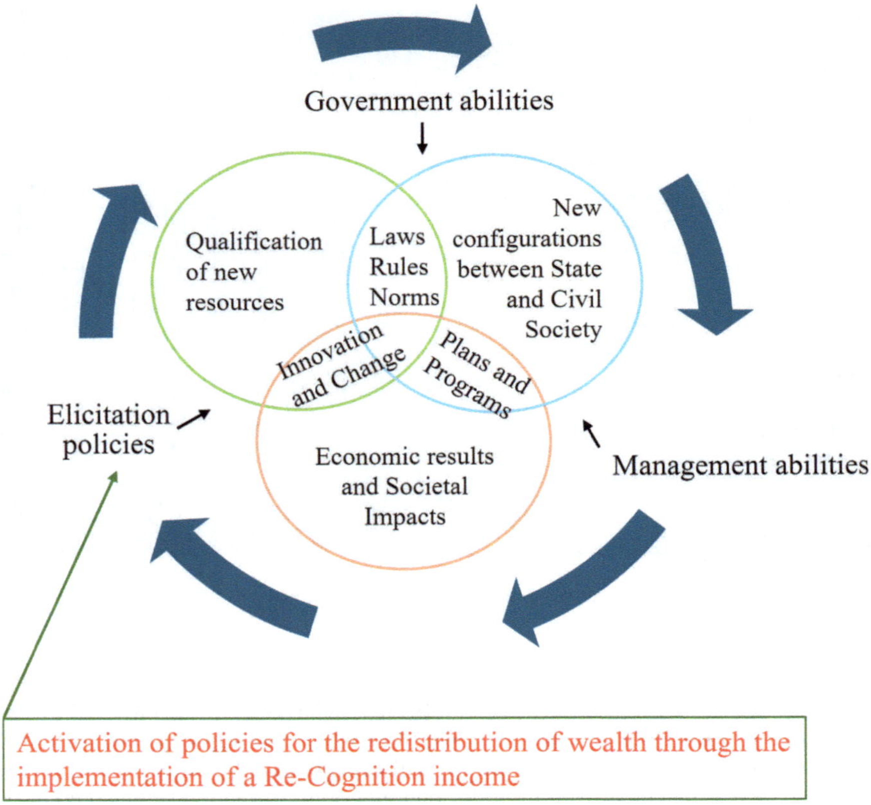

Fig. 11.8 Implementing Re_cognition Income. *Source* Authors' elaboration

activity at any level of a single process. Consider the creation of a public-owned platform (perhaps called Europe Online) enabling economic operators currently on Amazon to channel the commercial margin, now retained by a private entity, to their country's treasury, prioritizing distribution to various users who interacted with the platform.

Implications and Conclusions

Following the analysis of the evolving paradigms from Proprietary to Financial Capitalism and the emergence of Participatory Capitalism, it becomes evident that our society is at a critical juncture. The unprecedented pace of technological advancement and the resultant socio-economic transformations necessitate a reevaluation of how we govern and distribute wealth (Brynjolfsson & McAfee 2014; Schwab

2017; Harari 2018). As we transition into this new era, it is imperative to ensure that the benefits of technological progress are equitably shared and that the potential for social justice is realized.

These discussions are especially pertinent within the context of *Complexity and Emergence in Market Ecosystems*, where interconnected systems give rise to unpredictable outcomes, making governance and regulation both challenging and essential (Vargo et al. 2023). By examining these complex systems, we can better understand how emergent patterns of inequality and opportunity unfold in the digital economy.

A return to political primacy is urgent so that the decision-makers elected by the people can initiate a new era where the central factor is 'participation' in processes, rather than 'work' as previously understood, balancing wealth flows between 'World 1' and 'World 2' (Fig. 11.9). Achieving such a result requires reflecting on how, in our country even more than in many other Western nations, state bureaucracies have exercised completely autonomous activities, sometimes opposing executive branches, not only visibly adopting opposition positions but also, more dangerously, risking becoming a 'counterforce' of a corporate nature, effectively a political class. In a market ecosystem where complexity and interdependence drive emergent dynamics, reclaiming political oversight is vital to ensure that wealth and opportunities are distributed in ways that benefit society as a whole, rather than a select few.

In recent years, the growing autonomy of state bureaucracies has manifested in ways that challenge the intended roles of democratic governance. This self-directed operation of bureaucratic institutions has, in many instances, diverged from the priorities and directives of elected officials, leading to a form of governance that operates

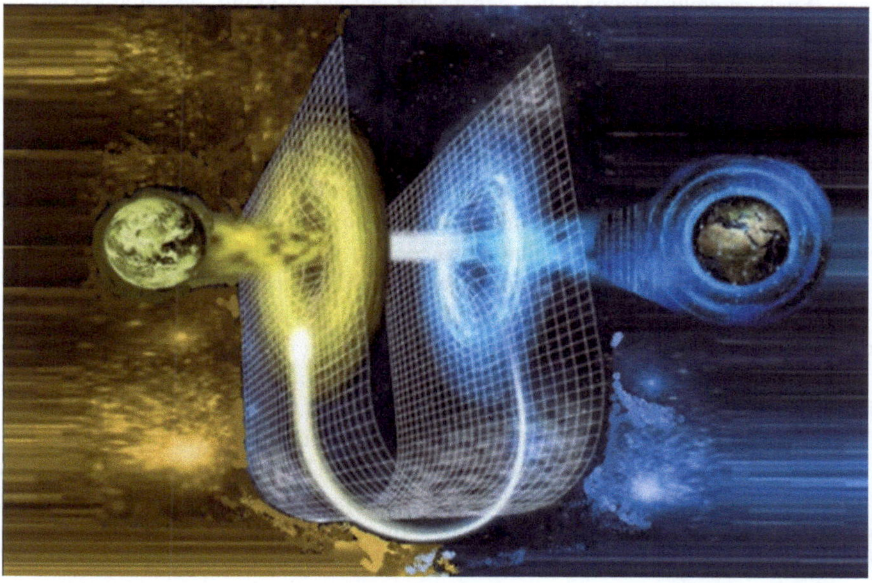

Fig. 11.9 Re_cognition towards 'World 1'. *Source* Authors' elaboration

independently of, and sometimes in opposition to, the political leadership. Such dynamics not only undermine the principle of representative democracy but also foster a scenario where bureaucracies effectively function as a separate political class, exerting significant influence over public policy and administration (Schmitt 2005; Fukuyama 2011; Habermas 1996).

With the advancement of modern technologies, particularly artificial intelligence (AI), this bureaucratic power is set to become even more entrenched. The delegation of decision-making to machines and algorithms threatens to push society towards a form of totalitarianism where human oversight and judgment are increasingly marginalized. In such a scenario, a depoliticized society would be at the mercy of anonymous algorithms, whose decisions lack transparency and accountability. This trend towards algorithmic governance raises significant ethical and practical concerns about the role of technology in shaping societal outcomes. From a market ecosystem perspective, the concentration of decision-making within algorithmic systems can lead to unintended consequences, where opaque and rigid processes disrupt natural market dynamics and stifle human agency. Ensuring that these systems remain accountable and adaptable is crucial for the healthy emergence of market ecosystems.

The new 'political classes' today tend to acquire and maintain power primarily through bureaucratic elements, supported by a kind of scientific metaphysics, as noted by Carl Schmitt. This involves a blind faith in pseudo-positive law, where abstract statements replace common sense, and self-referential ethical codes supplant established legal principles. This paradigm shift has led to a rigid intolerance towards dissenting or differing opinions, often cloaked under the guise of 'political correctness.' Such an environment stifles debate and critical thinking, which are essential for a healthy democratic process (Zuboff 2019; Pasquale 2015; O'Neil 2016).

The implications of this shift are profound. If left unchecked, the increasing dominance of technocratic governance and the erosion of political primacy could lead to a society where economic disparities are further entrenched, and social mobility is restricted. However, this technological evolution also presents opportunities to redefine work and productivity (Ford 2015; Mason 2015; Susskind and Susskind 2015). By examining these dynamics through the lens of market ecosystems, we gain insight into how technological advances can drive both positive and negative emergent effects within socio-economic systems. Recognizing and addressing these complexities is essential for promoting adaptive and resilient policies that benefit all members of society.

In a future where repetitive and mundane tasks are automated, humans could be liberated to engage in more creative, design-oriented, and intellectual endeavors. This shift could foster an environment where innovation thrives, and individuals are empowered to pursue their passions and talents. The potential for technology to enhance human creativity and problem-solving capabilities is immense, provided it is harnessed and directed with inclusive and ethical considerations in mind (Bostrom 2014; Tegmark 2017; Floridi 2014).

In conclusion, we must urgently reconsider the balance between the emerging digital realm ('World 2') and the material world ('World 1') to ensure equitable wealth distribution, recognizing the indispensable contributions of all participants, and addressing the profound socio-economic transformations driven by rapid technological advancements. It is crucial to restore the primacy of politics, ensuring that democratically elected leaders have the authority and capacity to guide and regulate the integration of technology in ways that serve the common good. By doing so, we can create a more inclusive and participatory society where technological progress benefits all, rather than exacerbating existing inequalities. This approach not only addresses the immediate challenges but also paves the way for a future where technology is a tool for human liberation and enrichment, enabling a society that values creativity, critical thinking, and equitable growth.

References

Anderson, B.: Imagined communities: reflections on the origin and spread of nationalism. Verso (1983)

Arrighi, G.: The long twentieth century: money, power, and the origins of our times. Verso (1994)

Arthur, W.B.: The nature of technology: what it is and how it evolves. Free Press (2017)

Barile, S., Saviano, M., Caputo, F.: A service research contribution to the global challenge of sustainability. J. Serv. Theory Pract. **27**(5), 1059–1081 (2017a)

Barile, S., Polese, F., Calabrese, M., Iandolo, F., Carrubbo, L.: A theoretical framework for measuring value creation based on the viable systems approach (vSa). In: Barile S. (ed.) Contributions to Theoretical and Practical Advances in Management. A Viable Systems Approach (VSA) (pp. 61–94). Roma: Aracne (2013)

Barile, S., Saviano, M., Iandolo, F., Caputo, F.: La dinamica della sostenibilità tra vortici e correnti: un modello a Tripla Elica. In: Sviluppo Sostenibilità e Competitività Delle Aziende. Il contributo degli economisti aziendali. Il Mulino (2017a)

Bauman, Z.: Liquid modernity. Polity Press (2000)

Bell, D.: The coming of post-industrial society: a venture in social forecasting. Basic Books (1973)

Benkler, Y.: The wealth of networks: how social production transforms markets and freedom. Yale University Press (2006)

Bodin, J.: On sovereignty: four chapters from the six books of the commonwealth. Cambridge University Press (1992)

Bostrom, N.: Superintelligence: paths, dangers. Oxford University Press, Strategies (2014)

Braverman, H.: Labor and monopoly capital: the degradation of work in the twentieth century. Monthly Review Press (1974)

Bryan, D., Rafferty, M.: Capitalism with derivatives: a political economy of financial derivatives. Palgrave Macmillan, Capital and Class (2006)

Brynjolfsson, E., McAfee, A.: The second machine age: work, progress, and prosperity in a time of brilliant technologies. W.W. Norton & Company (2014)

Castells, M.: The rise of the network society. Wiley-Blackwell (2010)

Chandler, A.D.: The visible hand: the managerial revolution in American business. Harvard University Press (1977)

Chesbrough, H.W.: Open innovation: the new imperative for creating and profiting from technology. Harvard Business School Press (2003)

Davenport, T.H., Prusak, L.: Working knowledge: how organizations manage what they know. Harvard Business School Press (2000)

Davis, G.F.: Managed by the markets: how finance reshaped America. Oxford University Press (2009)

Deleuze, G., Guattari, F.: A thousand plateaus: capitalism and schizophrenia. University of Minnesota Press (1987)
Van Dijck, J., Poell, T., De Waal, M.: The platform society: public values in a connective world. Oxford University Press (2018)
Drucker, P.F.: Post-capitalist society. HarperBusiness (1993)
Ferguson, T.: Golden rule: the investment theory of party competition and the logic of money-driven political systems. University of Chicago Press (1995)
Floridi, L.: The fourth revolution: how the Infosphere is reshaping human reality. Oxford University Press (2014)
Ford, M.: Rise of the robots: technology and the threat of a jobless future. Basic Books (2015)
Foucault, M.: Discipline and punish: the birth of the prison. Vintage Books (1995)
Friedman, T.L.: The world is flat: a brief history of the twenty-first century. Farrar, Straus and Giroux (2005)
Fukuyama, F.: The origins of political order: from prehuman times to the French revolution. Farrar, Straus and Giroux (2011)
Gaddis, J.L.: The cold war: a new history. Penguin Press (2005)
Golinelli, G.M., Barile, S., Spohrer, J., Bassano, C.: The evolving dynamics of service co-creation in a viable systems perspective. In: The 13th Toulon-Verona Conference Proceedings of the International Conference in Coimbra, Portugal (pp. 2–4) (2010)
Gowan, P.: The global gamble: Washington's Faustian bid for world dominance. Verso (1999)
Graeber, D.: Debt: the first 5000 years. Melville House (2011)
Habermas, J.: The theory of communicative action: reason and the rationalization of society. Beacon Press (1984)
Habermas, J.: Between facts and norms: contributions to a discourse theory of law and democracy. MIT Press (1996)
Habermas, J.: The postnational constellation: political essays. MIT Press (1998)
Hacker, J.S., Pierson, P.: Winner-take-all politics: how Washington made the rich richer--and turned its back on the middle class. Simon (2010)
Harari, Y.N.: 21 Lessons for the 21st century. Spiegel & Grau (2018)
Harvey, D.: A brief history of neoliberalism. Oxford University Press (2005)
Helleiner, E.: States and the Reemergence of global finance: from Bretton woods to the 1990s. Cornell University Press (1994)
Hobsbawm, E.J.: The age of empire: 1875–1914. Weidenfeld & Nicolson (1987)
Hobsbawm, E.J.: Industry and empire: the birth of the industrial revolution. New Press (1999)
Judt, T.: Postwar: a history of Europe since 1945. Penguin Press (2005)
Klein, N.: The shock doctrine: the rise of disaster capitalism. Picador (2008)
Krippner, G.R.: Capitalizing on crisis: the political origins of the rise of finance. Harvard University Press (2011)
Landes, D.S.: The unbound Prometheus: technological change and industrial development in western Europe from 1750 to the Present. Cambridge University Press (1969)
Langley, P.: The everyday life of global finance: saving and borrowing in Anglo-America. Oxford University Press (2008)
Lazonick, W., O'Sullivan, M.: Maximizing shareholder value: a new ideology for corporate governance. Econ. Soc. **29**(1), 13–35 (2000)
LiPuma, E., Lee, B.: Financial derivatives and the globalization of risk. Duke University Press (2004)
Lury, C.: Consumer culture. Polity Press (2011)
Lusch, R., Vargo, S., Mele, C., Polese, F.: Service-dominant logic. premesse, prospettive, possibilità. Wolters Kluwer CEDAM (2017)
MacKenzie, D.: An engine. How Financial Models Shape Markets. MIT Press, Not a Camera (2006)
Marx, K.: Capital: a critique of political economy. Penguin Classics (1992)
Mason, P.: PostCapitalism: a guide to our future. Farrar, Straus and Giroux (2015)

Mokyr, J.: The lever of riches: technological creativity and economic progress. Oxford University Press (1990)
Nonaka, I., Takeuchi, H.: The knowledge-creating company: how Japanese companies create the dynamics of innovation. Oxford University Press (1995)
O'Neil, C.: Weapons of math destruction: how big data increases inequality and threatens democracy. Crown Publishing Group (2016)
Pasquale, F.: The black box society: the secret algorithms that control money and information. Harvard University Press (2015)
Piketty, T.: Capital in the twenty-first century. Harvard University Press (2014)
Polanyi, M.: The Tacit Dimension. Doubleday & Company (1966)
Polanyi, K.: The great transformation: the political and economic origins of our time. Beacon Press (2001)
Polese, F., Sarno, D., Troisi, O., Grimaldi, M.: From B2B to A4A: an integrated framework for viable value co-creation. Mercati & Competitività **2018**(3), 135–161 (2018). https://doi.org/10.3280/mc2018-003008
Poon, M.: From new deal institutions to capital markets: commercial consumer risk scores and the making of subprime mortgage finance. Acc. Organ. Soc. **34**(5), 654–674 (2009)
Porter, M.E.: Competitive advantage: creating and sustaining superior performance. Free Press (1985)
Rajan, R.G.: Fault lines: how hidden fractures still threaten the world economy. Princeton University Press (2010)
Rifkin, J.: The zero marginal cost society: the internet of things, the collaborative commons, and the eclipse of capitalism. Palgrave Macmillan (2014)
Rodrik, D.: The globalization paradox: democracy and the future of the world economy. W.W. Norton & Company (2011)
Russo, J.A.: Capital, crisis, and class struggle: the future of work in the twenty-first century. Monthly Review Press (2019)
Schmitt, C.: The concept of the political. University of Chicago Press (2007)
Schmitt, C.: Political theology: four chapters on the concept of sovereignty. University of Chicago Press (2005)
Schumpeter, J.A.: Capitalism, socialism and democracy. Harper Perennial (2008)
Schwab, K.: The fourth industrial revolution. Crown Business (2017)
Sennett, R.: The culture of the new capitalism. Yale University Press (2006)
Srnicek, N.: Platform capitalism. Polity Press (2017)
Stiglitz, J.E.: Globalization and its discontents. W.W. Norton & Company (2002)
Stiglitz, J.E.: Freefall: America, free markets, and the sinking of the world economy. W.W. Norton & Company (2010)
Stiglitz, J.E.: The price of inequality: how today's divided society endangers our future. W.W. Norton & Company (2012)
Susskind, R., Susskind, D.: The future of the professions: how technology will transform the work of human experts. Oxford University Press (2015)
Tapscott, D.: The digital economy: rethinking promise and peril in the age of networked intelligence. McGraw-Hill (2015)
Teece, D.J.: Dynamic capabilities and strategic management: organizing for innovation and growth. Oxford University Press (2009)
Tegmark, M.: Life 3.0: Being human in the age of artificial intelligence. Knopf (2017)
Thompson, E.P.: The making of the English working class. Victor Gollancz Ltd (1963)
Tilly, C.: The formation of national states in western Europe. Princeton University Press (1975)
Tilly, C.: Coercion, capital, and European states, AD 990–1990. Blackwell (1990)
Tocqueville, A.: Democracy in America. Penguin Classics (2003)
Tooze, A.: Crashed: how a decade of financial crises changed the world. Viking (2018)
Vargo, S.L., Lusch, R.F.: Evolving to a new dominant logic for marketing. J. Mark. **68**(1), 1–17 (2004)

Vargo, S.L., Lusch, R.F.: Service-dominant logic: continuing the evolution. J. Acad. Mark. Sci. **36**, 1–10 (2008)

Vargo, S.L., Peters, L., Kjellberg, H., Koskela-Huotari, K., Nenonen, S., Polese, F., Sarno, D., Vaughan, C.: Emergence in marketing: an institutional and ecosystem framework. J. Acad. Mark. Sci. **51**(1), 2–22 (2023)

Weber, M.: The Protestant ethic and the spirit of capitalism. Allen and Unwin (1930)

Zuboff, S.: The age of surveillance capitalism: the fight for a human future at the new frontier of power. PublicAffairs (2019)

Sergio Barile is a Full Professor of Economics and Business Man-agement (SECS-P/08) at Sapienza University of Rome. He has served as Dean of the Department of Management and as Coordinator of the PhD Program in Man-agement, Banking, and Commodity Sciences at Sapienza University. His research interests encompass gov-ernment and business management, systems theory, decision theory, and complexity. Professor Barile is the founder of the Association for Research on Viable Sys-tems (ASVSA) and is widely recognized as a key contributor to the development of the Viable Systems Ap-proach (vSa). He serves on the editorial boards of several specialized journals in business and economic man-agement and has authored numerous books and articles published in both national and international journals. In 2015, he was honored with the Evert Gummesson Out-standing Research Award.

Francesca Iandolo PhD, is an Associate Professor of Management (SSD: SECS-P/08, SC: 13/B2) at the Department of Management, Sapienza University of Rome. She teaches under-graduate courses in Business Economics and Management and master's level courses in Management of In-novation and Artificial Intelligence. She is also the Coordinator for the Master's Degree in Economics and Communication for Management and Innovation. Her research explores the application of systems theories to corporate sustainability and the role of technological diffusion within digital platforms and innovation eco-systems. Dr. Iandolo has actively participated in nu-merous national and international conferences as a dis-cussant and has published widely in both national and international journals. She serves as an ad-hoc reviewer for prominent international journals in management, innovation, and corporate sustainability and has been the Principal Investigator for various national and in-ternational projects, including the coordination of an Erasmus+ project. In 2020 and 2023, she was a Visiting Researcher at the School of Business and Management, Queen Mary University of London. Since 2021, Dr. Iandolo has served as the Responsible Academic Supervisor for International Mobility Agreements within the Faculty of Economics at Sapienza University of Rome, overseeing international mobility for courses in the Department of Management. She also represented researchers on the Departmental and Faculty Councils from 2020 to 2024. Additionally, she has been a licensed Chartered Accountant and Auditor since 2011.

MIX
Papier aus verantwortungsvollen Quellen
Paper from responsible sources
FSC® C105338

If you have any concerns about our products,
you can contact us on
ProductSafety@springernature.com

In case Publisher is established outside the EU,
the EU authorized representative is:
Springer Nature Customer Service Center GmbH
Europaplatz 3, 69115 Heidelberg, Germany

Printed by Libri Plureos GmbH
in Hamburg, Germany